T0074035

PRAISE FOR *END OF ABUNDANCE IN TECH*

"For any senior executive, this book will become an outstanding and trusted reference for determining effective approaches to ensure value is captured from the increasing volumes of data that we all seem to be drowning in. Ben DeBow's *End of Abundance in Tech* is a practical guide to ensuring value is realized from nearly any IT environment in a realistic and simplified way that exceeds any previous attempts."

JASON CHILD, CFO of Arm Limited; Board member at Coupang;
former CFO of Splunk, Opendoor, and Groupon

"I found myself nodding in agreement with many of the scenarios presented in *End of Abundance in Tech* and was impressed with the practical and actionable advice offered. This book provides insightful and thought-provoking perspectives on enhancing efficiencies in technology. Ben DeBow effectively uses real-life examples and situations to illustrate his points, making it easier for the reader to relate."

TUSHAR THAKKER, VP Product Development,
Quickbooks Payroll & Time

"Ben DeBow's book, *End of Abundance in Tech*, will benefit any IT, business, or data leader looking to develop a successful data and analytics strategy. Via solid research and numerous illustrations, DeBow draws an intriguing anaogy between personal health and technology health and provides practical advice on how to find efficiencies to drive business value."

DOUGLAS LANEY, Author of *Infonomics* and *Data Juice;*
Innovation Fellow with West Monroe Partners;
Guest lecturer on the valuation of information, MIT

"Ben DeBow is at the leading edge of the next wave of business innovation. While many organizations are trying to mature their ability to capture and leverage data to gain new insights, *End of Abundance in Tech* gives leaders in any organization a framework to connect business value and technology investment. A must read for business leaders seeking to improve the efficiency and effectiveness of their tech!"

JEREMIAH HURLEY, Senior Associate in Booz Allen Hamilton's Chief Technology Office; Former Deputy CTO and Director of Innovation at JSOC, US Department of Defense

"In thirty years of working in tech infrastructure and information security, I haven't encountered a person who can identify and solve workload performance issues the way Ben DeBow does. In *End of Abundance in Tech*, he shows you that you *can* gain the insight you need to run highly optimal workloads and truly understand your costs. This book is a must-read for any technology executive chasing the elusive big picture of how their environment performs and what it really costs. While we all may know that "non-optimal" almost translate into a more enormous bill, we don't always know why. Even if you think you've got these answers down pretty well, you'll come away with new ways to think about and solve issues for the long term — not just this week. You may even be able to finally end the debate in your shop about whether it's the infrastructure that's the problem or that less-than-perfect code."

JUSTIN M. CUYLER, Sr Director of Information Security, WellSky Corp; Former Sr. Director of Security and Compliance, CarePort Health

"In *End of Abundance in Tech*, Ben DeBow conveys wisdom and insight into the challenges and opportunities of managing modern enterprise systems. Clearly, the author knows how to optimize workload performance for technology efficiency, as illustrated by the practical guidance interspersed with real-life examples and astute observations. Using DeBow's strategies, readers will gain new ways of thinking about system health, operational efficiency, and financial transparency in technology."

DAVID HYDE, CTO Vertical Markets Software Solutions, a division of Global Payments

"End of Abundance in Tech is a valuable resource for any IT leader looking to have a measurable and meaningful impact on the business. Ben DeBow presents a simple argument for gearing technology towards improving efficiency and driving business value. The book provides practical strategies from DeBow's firsthand cross-industry experiences of achieving cost savings, driving efficiency, and improving productivity, while still delivering value to the business. I would recommend this book to anyone in the IT business looking for actionable solutions to impact their bottom line and stay ahead of the curve in the ever-dynamic business environment of today."

STEVE WANAMAKER, CEO and Publisher, *CDO Magazine*

END OF
ABUNDANCE
IN TECH

END OF ABUNDANCE
IN TECH

HOW IT LEADERS CAN FIND EFFICIENCIES

TO DRIVE BUSINESS VALUE

BEN DEBOW

Forbes | Books

Published by Forbes Books, Charleston, South Carolina.
Member of Advantage Media.

Forbes Books is a registered trademark, and the Forbes Books colophon is a trademark of Forbes Media, LLC.

Printed in the United States of America.

10 9 8 7 6 5 4 3 2 1

ISBN: 979-8-88750-032-4 (Hardcover)
ISBN: 979-8-88750-033-1 (eBook)

LCCN: 2023904265

Cover design by Megan Elger.
Layout design by Lance Buckley.

This custom publication is intended to provide accurate information and the opinions of the author in regard to the subject matter covered. It is sold with the understanding that the publisher, Forbes Books, is not engaged in rendering legal, financial, or professional services of any kind. If legal advice or other expert assistance is required, the reader is advised to seek the services of a competent professional.

Since 1917, Forbes has remained steadfast in its mission to serve as the defining voice of entrepreneurial capitalism. Forbes Books, launched in 2016 through a partnership with Advantage Media, furthers that aim by helping business and thought leaders bring their stories, passion, and knowledge to the forefront in custom books. Opinions expressed by Forbes Books authors are their own. To be considered for publication, please visit **books.Forbes.com**.

For my mother for showing me what's possible.

C O N T E N T S

INTRODUCTION

How healthy are you? If you asked four different doctors or *specialists*, this question might get you four different answers. Your oncologist might look at your MRI scan and say, "You are cancer-free."

Your orthopedist might look at your X-ray and say, "Your knee is perfectly healed."

Your dentist might tap on a problem tooth and say, "You need a root canal."

Your cardiologist might look at your EKG and say, "You've had a mild heart attack."

Or you may get a glowing prognosis for your health today, but that doesn't mean you're ready to run a marathon tomorrow. You would have to prepare and make sure you have enough lung *capacity* to support such a change in your *system*'s regular routine or *processing load*. You can't be expected to run 26.2 miles just because you are deemed "healthy."

The same idea holds true for the health of your technology environment. So how healthy is your technology environment?

Your chief data officer (CDO), chief information officer (CIO), chief technology officer (CTO), and chief financial officer (CFO) are all specialists in their own right and, as such, might each interpret this question differently. (Chances are your chief executive officer, or CEO, would steer you to one of the above if asked.) They may each look at different information, or *data*, to make this assessment. And they will likely make a conclusion based on their own area of expertise using some tool or solution to analyze the data before them.

The truth is no one really knows for sure how healthy his or her technology environment is today and how much it will be able to support tomorrow. All they know is what they see in the large quantities of data being collected as technology systems process more and more business transactions and data every day.

I've seen this situation play out countless times in conversations with executives from Silicon Valley start-ups to leading insurers, healthcare and financial institutions, retailers, and more. Everyone views the health of a system based on their own personal stake in it. Regarding the health of your body, the answer you get today might be different from the one you get three months from now. The same holds true for the health of your entire business, which relies on your technology environment. Yet the way we measure the health of technology hasn't really changed in twenty years.

THE ERA OF ABUNDANCE IN TECHNOLOGY

Organizations have been spending increasingly large amounts of money on technology over the past twenty or so years. From 2005 to 2023, worldwide technology spending is projected to reach nearly $5 trillion,[1] a more than 81 percent increase, according to Gartner. From 2022 to 2023 alone, an increase of 5.1 percent is expected (compared

---

1 Statista, "Information Technology (IT) Worldwide Spending from 2005 to 2023," August 16, 2022, https://www.statista.com/statistics/203935/overall-it-spending-worldwide.

to a 2021/22 increase of 3 percent), as "demand for technology in 2023 is expected to be strong as enterprises push forward with digital business initiatives in response to economic turmoil."[2]

Information Technology (IT) worldwide spending from 2005 to 2023

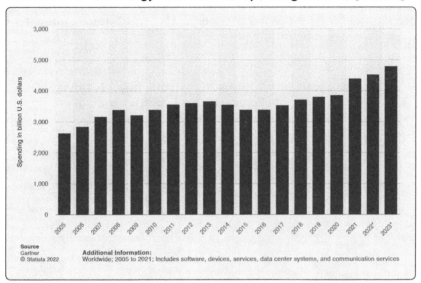

Figure 1: Global IT worldwide spending, 2005–2023.
Source: Gartner as reported by Statista 2022
(https://www.statista.com/statistics/203935/overall-it-spending-worldwide)

IDC predicted that the world's technology buyers would spend more on infrastructure intended for use in clouds than in other scenarios during 2022 and that cloud infrastructure would exceed $90 billion, or 22 percent growth, compared to 2021. Shared cloud infrastructure was predicted to grow by 24.3 percent to $63.9 billion for the full year, while dedicated cloud infrastructure was projected to improve sales by 16.8 percent to $26.3 billion for the full year.

---

2   Gartner, "Gartner Forecasts Worldwide IT Spending to Grow 5.1% in 2023," October 19, 2022, https://www.gartner.com/en/newsroom/press-releases/2022-10-19-gart-ner-forecasts-worldwide-it-spending-to-grow-5-percent-in-2023.

Noncloud infrastructure was predicted to grow just 1.8 percent to $60.7 billion.[3]

Long term, IDC predicts spending on cloud infrastructure will have a compound annual growth rate of 12 percent over the 2021–2026 forecast period, reaching $134 billion in 2026 and accounting for 67.9 percent of total compute and storage infrastructure spend.[4]

Worldwide Enterprise Infrastructure Buyer & Cloud Deployment Forecast, 2021 - 2026 (Spending)

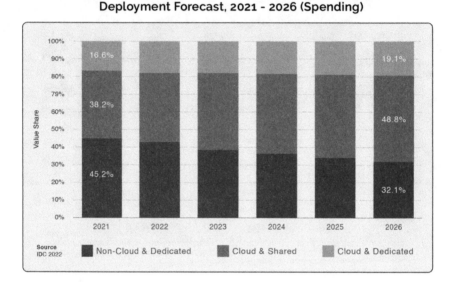

Figure 2: Worldwide enterprise infrastructure buyer and cloud deployment forecast, 2021–2026. Source: IDC 2022 (https://www.idc.com/getdoc.jsp?containerId=prUS49732022)

3   Simon Sharwood, "Cloud Infrastructure Spend to Crack $90b and Overtake Non-Cloud in 2022," *Register*, July 4, 2022, https://www.theregister.com/2022/07/04/idc_cloud_spend_predictions.

4   IDC, "IDC Tracker Finds Spending on Compute and Storage Cloud Infrastructure Increased Strongly across Most Regions in the Second Quarter of 2022," press release, accessed from Business Wire, September 30, 2022, https://www.businesswire.com/news/home/20220930005135/en/IDC-Tracker-Finds-Spending-on-Compute-and-Storage-Cloud-Infrastructure-Increased-Strongly-Across-Most-Regions-in-the-Second-Quarter-of-2022.

Figure 2 provides a visual illustration of what I refer to as the era of abundance in technology, largely stemming from the shift away from noncloud to cloud-based services.

According to Sid Nag, vice president analyst at Gartner:[5]

Cloud computing will continue to be a bastion of safety and innovation, supporting growth during uncertain times due to its agile, elastic and scalable nature. Yet, organizations can only spend what they can spend.

Nag further goes on to say:

Cloud spending will continue through perpetual cloud usage. Once applications and workloads move to the cloud they generally stay there, and subscription models ensure that spending will continue through the term of the contract and most likely well beyond. For these vendors, cloud spending is an annuity—the gift that keeps on giving.

As organizations continue to migrate to the cloud, these bills are only going to go up, perpetually funding these vendors' "annuities." It's my belief that technology leaders today don't fully comprehend the Total Cost of Ownership (TCO) of all the different technology platforms they're migrating or purchasing, especially when they go to a utility, or *pay-for-what-you-use*, model, which is how they pay for compute, or software, in the cloud. Conventional wisdom dictates, "You have to be in the cloud," but then there is sticker shock once you get there.

---

5   Sid Nag, "Gartner Forecasts Worldwide Public Cloud End-User Spending to Reach Nearly $600 Billion in 2023," Gartner, October 31, 2022, https://www.gartner.com/en/newsroom/press-releases/2022-10-31-gartner-forecasts-worldwide-public-cloud-end-user-spending-to-reach-nearly-600-billion-in-2023.

Public cloud services end-user spending worldwide from 2017 to 2023 (in billion U.S. dollars)

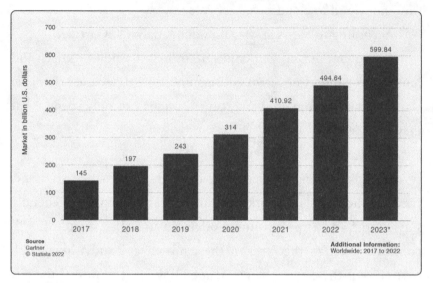

Figure 3: Public cloud services end-user spending worldwide, 2017–2023. Source: Gartner as reported by Statista 2022 (https://www.statista.com/ statistics/273818/global-revenue-generated-with-cloud-computing-since-2009)

The growth rate for public cloud services end-user spending is increasing, from 20.3 percent between 2021 and 2022 to 21.3 percent between 2022 and 2023 (projected), according to Gartner.[6] This spending growth curve since 2017 is likely to disproportionately keep increasing were we to project this chart out over the next three to five years. Welcome to the era of abundance in technology.

The only way organizations are dealing with these increased costs today is to reduce operational costs, such as the number of servers or what is stored on the servers they have. But if your business is successful—and that is the goal—you will be amassing new customers and

---

6   Statista, "Public Cloud Services End-User Spending Worldwide from 2017 to 2023," April 20, 2022, https://www.statista.com/statistics/273818/ global-revenue-generated-with-cloud-computing-since-2009.

more transactions, which means you will have more data, which means you will need more storage, which means you will need those servers unless you optimize your applications and processes. Long term, the cloud will continue to offer more features as part of the service, but clients will have to pay a premium for that. This is called PaaS, or platform as a service, a division of cloud computing that spares clients the complexity of building and maintaining their own infrastructure within which to develop and launch applications.

There is another option. Understanding your workload analytics and identifying inefficiencies within your technology platforms—mostly around code and processes—can save enterprises millions, if not tens of millions, of dollars. These savings can then be reallocated back to the business and used to find ways to deliver more value.

Most CIOs, IT directors, and VPs think about costs. Some think about technology health

Understanding your workload analytics and identifying inefficiencies within your technology platforms—mostly around code and processes—can save enterprises millions, if not tens of millions, of dollars.

and leverage many of the tools in the marketplace to gauge that today. Very few are thinking about technology efficiency. And the way we get to efficiency is by having financial transparency into our true costs within the applications and services so we understand where the costliest processes are in the enterprise. This is how we put an end to the era of abundance in technology and why I wrote this book.

OPTIMIZING TECHNOLOGY HEALTH

This deficiency in enterprises knowing the health, efficiency, and financial impact of their technology and what actions they need to take next led me to start my own company to help business leaders solve problems by ensuring that their systems are equipped to meet the needs of the company overall. Further, I maintain there are inter-dependencies between technology and financial performance that are not fully understood and therefore not fully optimized in most organizations. My company, Fortified, helps executives see the value in their technology spending and understand how they can save money and resources by ensuring their technology systems are healthy, efficient, and stable to meet the workload of today and the growth of tomorrow.

I founded Fortified Data, a next-generation database consultancy focused on planning and designing data systems for performance and scalability, on January 1, 2002, with a vision to provide businesses with innovative ways to discover the value of their data and scale their systems to meet their business goals and objectives while reducing risk. We take a 360-degree, broad view of workload health. In 2022 we rebranded as Fortified, recognizing the potential of our services to transform business operations on a greater scale.

I hold academic degrees in accounting and information systems from the University of Cincinnati. Data was a part of my studies, but it was only by being exposed to businesses as a consultant and information architect in the late 1990s that I realized the potency of data. It made sense to me, and it became my passion. My accounting background gave me a unique lens through which to view data that wasn't very common at the time, one of financial transparency and the value equation. If this graph line goes up, how much does it cost the company? If it goes down, how much does it save? This is why

I urge CFOs of businesses today to claim their rightful stake in the discussion of technology health as well.

Every business is bringing on more technology services to the point where companies that make products suddenly find themselves in the business of technology as well. Take Target, for example, and how they were able to scale their brick-and-mortar empire during the COVID-19 pandemic through technology. Now they're also in the business of digital experiences—order online, pick up in store, purchase through the app, etc. "Target has taken the friction out of the shopping process, making it as easy as possible for customers to do business with them," according to Michael Lasser, equity analyst at UBS.[7] We, as technology professionals, need to take the friction out of data performance management as well.

DATA GROWTH IS DRIVING TECHNOLOGY INVESTMENT

What is data? Data is essentially any information you'd want to collect or store that describes the state or value of something. Everything in technology is done to be able to store, move, analyze, or present data for people. Every app you use performs a function—it's an organized way for you to see data and make a decision in regard to a transaction. Twenty years ago data was just the information you needed to run a business. Today everybody views data and wants to understand the information that's going to make their lives (or their businesses) better, more valuable, more strategic, or whatever their goal or objective is. The goals that are your priority will inform the areas you focus on first.

Data used to be strictly a means to an end. But today data is a foundational component in life.

---

7   Kelly Ell, "Target Is Winning the Pandemic Race," *WWD*, February 2, 2021, https://wwd.com/business-news/business-features/target-pandemic-retail-winner-1234719937.

In my view there are three key drivers elevating the focus on the health, efficiency, and financial impact of technology to a new level of importance in the early 2020s and beyond:

1. **Scale** - The volume of data is only growing as things get bigger, faster, and more immediate in the always-on, real-time virtual environment in which we live. Solving for scale means (1) how you make your system more efficient so that you can better leverage the existing technology resources you have and (2) how you make your system more mature to enable it to meet higher volume demands.

2. **Value** - Managing and maintaining your system within this environment requires having more clarity on what you need and how much it costs and knowing what value this effort brings—value not just to the company in terms of the bottom line but to the customer or end user as well.

3. **Health** - More business transactions and data require additional resources. Do you have enough resources to perform the functions that are necessary to operate your business? Are you experiencing over- or undercapacity? Is your server able to process business transactions while meeting the performance expectations of its users? Most people focus on having enough resources, but having more resources than you need takes away from the overall value equation and must be monitored as well. Understanding health and capacity also gives you an all-important view of the future. You're not just optimizing the system for today but you're also optimizing it for the years ahead.

The goal for CDOs, CTOs, and CIOs is to easily articulate successes and challenges to different layers of management in the company so that leadership can make informed decisions. Current solutions involve a lot of nice graphics and dashboards, but what's fundamentally missing are actual insights—even in mature companies, it's difficult to analyze technology data and translate it into insights that can be acted on in the marketplace.

Many new technology tools and solutions have arisen to give executives better access to performance and other technology metrics, but data visualizations only beget more data and reports, not insights and actions. Over the last twenty years, the tools to show data, insights, and highlights have improved drastically, but they fail to prescribe specific actions such as *change X now and Y next to achieve Z.* Insights are great, but they don't tell the technical folks what actions they need to take to make their systems better or what actions are most aligned with the key performance indicators (KPIs).

IS YOUR TECHNOLOGY ENVIRONMENT HEALTHY?

I've dedicated my career to answering this question for hundreds of businesses. The answer usually is, "It could be healthier" or "It's healthy today, but with future growth, the system will not be healthy or have enough capacity." Business leaders want to know what to do. They don't really care if one line on a chart is going up while another is going down.

The data we collect can only play a role in determining the health, efficiency, and financial impact of a system if it is translated into actions. Those actions must be mapped to KPIs or objectives and key results (OKRs) at the top level of the enterprise to help executives know if their systems are performing optimally and are delivering technology services that achieve the desired objectives. Unlike CFOs who have easy access to journal entries that can produce a P&L or balance sheet, business executives do not have an instant, holistic view into the health of the business at any time, leading to an information gap.

Most workers in any organization are task based. "What will it take to get me from here to there? To reach this KPI?" But from a leadership perspective, we need to ask different questions when looking at data:

- What actions can we take today that will make the most impact on the business tomorrow?
- What impact will it bring, and how do we capture the impact?
- What value can we derive from the change in financial value?

HEALTH, EFFICIENCY, AND THE FINANCIAL IMPACT OF TECHNOLOGY

We, as technology professionals, need to be able to provide intelligent actions with meaningful measurement, along with a financial view of the data. Then we need to map and measure as we're making changes to determine if our technology health is improving, and if so, how? And then we should layer the power of artificial intelligence (AI) and machine learning (ML) over those actions to refine the models and make smarter recommendations as to what needs to be changed in the future based on which of the changes already made have been the most successful.

My vision is to enable the science of workload analytics to include diving into every system that runs an enterprise and every computer that's in every data center or in the cloud and, for each one, understand the following:

- Is the server healthy?
- Is the server rightsized?
- Is the application workload efficient?
- What is the total cost of the applications today and in five years?

- What could the cost savings be if the application was 20 percent more efficient?

It starts by changing the way people think. I'm trying to articulate a new thought process for the technology industry. And that's what this book is about. I want to highlight, educate, and show businesses how to take the steps necessary to answer these questions—the right questions—to manage their environments differently.

One of the key tenets you will notice me repeat throughout this book is that while technology may have changed over the past twenty years—hello, GPS; hello, Alexis—people still look at the health of technology systems the way they did in the 1990s. And that's got to change.

If at the end of the day a business can say it has the right tools in place to measure the health of its systems in terms of productivity and financial impact and that its people understand and are doing the work that's needed now, then I will have realized my vision. I will have delivered because there'll be a definitive improvement in the systems, one that's instrumented. And the leaders in charge of these systems will see financial impact, as well as performance impact.

This book will show that the solution is to stop designing for complexity and instead focus on simplifying. It will help you—the CDO, CIO, CTO, CFO, and all business leaders—understand how to

- assess the health of your technology services,
- take action to maintain or improve the health of your system,
- identify where there are opportunities to reduce costs while optimizing performance, and
- bring financial transparency to your technology services.

There is a better way to simplify systems and capture, analyze, and use data to create and measure actions that will improve the maturity of a company and help it to scale.

I'll discuss steps you can take to define the health of your system and understand its value equation and how to interpret what actions to take to improve system availability, stability, performance, and capacity. I'll help you create a game plan to set in motion so that no matter the SME (subject matter expert), data observability, or application performance monitoring tools you use, you can create real impact on productivity and take control of your destiny when it comes to costs. By doing so, we can move beyond the era of abundance to the era of efficiency in technology. Are you with me?

C H A P T E R 1

There Is a Better Way

O ne of the most fascinating things about technology is you don't
know how much you need it until you have it. Take the smartphone, for example. Do you have any idea how you managed
life before having this six-inch computer in your pocket?

More specifically, take the fitness app on your smartphone in
your pocket. (Full disclosure: I am an avid hiker, sports enthusiast,
and adventure seeker, so many of my examples will reference human
health as well as technology health.) Ten years ago no one cared how
long or how well they slept, how far they ran, or how many steps they
walked in a day. Now someone may feel like a personal failure if he
or she doesn't meet their daily ten-thousand-step threshold or, what's
worse, loses out to a friend or family member who took more steps
that day than they did.

You may think the "ten thousand steps a day" rule was a brilliant
marketing play by smartphone manufacturers to get people to use
their phones' built-in fitness app. And it was brilliant marketing, but
it didn't originate with Apple or Samsung. It originated in 1965 with
a Japanese pedometer called Manpo-kei (literally translated as "ten-
thousand-step meter") for what might be considered the first modern
fitness app.

What recent interpreters of the "ten thousand steps a day" challenge seized on was not that there is any documented research declaring that ten thousand steps a day will make you healthier—there isn't. Rather, they knew that people are inherently goal oriented. Having a goal—in this case, ten thousand steps a day—simply makes you want to get up and start stepping.

> *Obstacles are those frightful things you see*
> *when you take your eyes off your goal.*
>
> **—HENRY FORD**

Why do you want to set a goal of ten thousand steps a day for yourself? To lose weight? To become healthier? To train for a marathon? The measurement itself doesn't mean much unless it's tied to a goal.

When you think about what we're doing as technologists, if we are supporting and optimizing technology systems and measuring their performance, we need to have a goal. And that goal needs to go beyond availability, which is essentially just keeping the system up and running. If an application shows great performance but has excess capacity, is it optimized? The challenge is that the capacity metric alone can't really tell you if the system is getting healthier. There is no definitive metric on server health or efficiency. If I make a bunch of changes, there is no single indicator, or number, that everyone can look at and know for sure that the server is healthier or more efficient.

Another piece of the puzzle is what we are measuring against, commonly referred to as KPIs. These don't exist for server health either. And lastly, there's no measurement of the financial impact or value of the system. Along with data on how the technology is performing, CFOs and business owners want to know where their money is going. In other words, if I am making improvements to the system, what are the financial impacts of those improvements? There must be a way to access these in real time, definitively, and from within the system itself.

Analysis of the health and performance of a server is necessary to ensure it is operating and processing transactions optimally. Ongoing workload analytics can lead to more efficient and productive use of server resources, optimize cloud costs, and provide increased system reliability and stability.

However, this analysis, in and of itself, often involves in-depth and complex monitoring of numerous technology and financial metrics over time. This complexity often prevents correct, consistent, and efficient analyses—particularly when less experienced subject matter experts are overseeing it. It's difficult to (1) diagnose whether a problem exists, (2) know if an improvement is needed, (3) define which performance metrics are important to a particular problem or improvement, (4) define the action or change needed to address the issue, and (5) understand how a particular problem or improvement changes over time.

It's interesting that we expect doctors to be 100 percent accurate when we did not design the human body, but when it comes to technology systems, which we did design, we do not have the same expectations. Part of this is due to the complexity of systems, but if you think about it, it really comes down to not having the right data and insights to solve problems accurately. For some reason this has become accepted in the technology industry, and this complacency is what I'm hoping to challenge.

THE COMPLEXITIES OF DETERMINING TECHNOLOGY HEALTH

There is a lot of information we can glean from a server's performance data. Which performance counters indicate health or capacity? At what time of the day is optimal to schedule processes? When is there resource contention, a.k.a. a traffic jam, that needs to be addressed?

The challenge is that while many infrastructure experts can interpret these performance metrics, most business, application, and database teams cannot, nor should they have to. The application and database teams know the codes or processes, but they do not know the impact of their code deployment on any of the lines in figure 4. This is one gap in technology that needs to be solved, especially as systems migrate to the cloud, and one of those lines could influence monthly costs.

Most dashboards today are open to interpretation, which leads to different outcomes or observations depending on who is reviewing them and what their focus is. The number of performance metrics often contributes to the complexity of analysis through information overload. It can be difficult to identify which performance metrics are

> **The challenge is that while many infrastructure experts can interpret these performance metrics, most business, application, and database teams cannot, nor should they have to.**

important and when, as well as the relation of one metric to another over time or in response to changes. The dashboards are not built for code analysis; there is no line to track that represents code. Now if you're in the technology trenches, this may sound complex and disheartening, but there is hope.

In our lives many of us have some type of app on our phone that makes it simple for us to identify actions we should take to improve our health, such as not eating late at night to improve our sleep score. This is done by collecting data about us all day, running in the background, so to speak, as we go about our lives. What if we had something like this for technology so that if we created actions that changed the data in our charts, we could have a simple view into the impact of these changes? We keep doing the actions that work; we try something else for those that don't.

Consider the performance image below from a Microsoft Windows server running SQL Server. If this image represents a company's database workload, would you say it is healthy? Without knowing what any of the lines or color gradations mean, would you say it is stable, reliable, and predictable? How much capacity does it have, and when should you schedule processes on this system? What actions could we take to change the data and cause our workload health to increase or decrease?

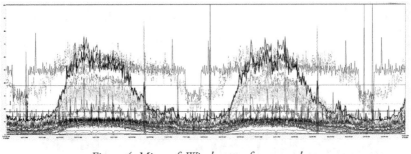

Figure 4: Microsoft Windows performance data.

Overall this system is optimized, but there are periods when there is resource pressure or the processes can be further optimized.

Figure 4 indicates some anomalies that need further investigation. If this represented a mission-critical system, one area of inconsistency (i.e., *friction*) could signal a real application problem or a drain on costs depending on what business processes were running during that time. It's not unlike the impact of unexpected traffic jams when you have a goal of getting somewhere on time; servers have traffic jams also. Friction costs time and money to alleviate, and the more frequent the friction, the more troublesome.

When your technology team needs to address the stability, reliability, and predictability of an application, they must understand which improvements will make the biggest impact and in what order to make them to achieve the biggest impact on the system. We need to identify what changes to make to the more complex systems (whether that's a change in configuration, timing, code, schema, or an index) so that our system can start to look cleaner. Today most application and database teams make system or application changes but do not know exactly what the impacts are because they are not experts on this data. Many times they don't even have access to this data, let alone know how to interpret it. And when we desire to reach that state of having an optimal "workload," how do we define it, and how do we communicate to the technology team that we have reached that state? This is even harder when you consider that enterprises have hundreds or thousands of applications with different KPIs that are all in different states of health.

INCREASING CAPACITY MAY NOT REDUCE FRICTION

Workload friction, as we refer to it, is contention within systems that requires more energy and impact response times. Friction means there is a traffic jam in the system, visually appearing as a big ball of thread on a dashboard, just like the red lines on Google Maps during rush hour. We need to research that occurrence and determine if the friction is due to resource contention, logical contention, or inefficient processes running at a nonoptimal time.

Resource contention means you don't have enough of something needed for the system to run properly. It could be memory, storage, or CPU (compute)—there is an insufficiency somewhere that is causing your system to run slowly. *Logical contention* is when processes are blocked or there are conflicts in the application. If multiple processes are running and need access to the same data, this could cause a blocking chain to occur, just as too many cars driving on the highway can cause a traffic jam. *Inefficient processes*, such as those resulting from poor code, can also cause a slowdown, negatively impacting an application. If you have any of these sources of friction, your system is not functioning optimally. Increasing capacity could be an answer, but we would first need to understand the source of the pain before taking any action.

Take this simple example of a traffic jam on a four-lane highway. There are logs spilled onto the road, with two cars approaching. This represents friction, a block, a challenge. You don't need to increase the capacity of the highway (read: add more lanes) because there are only two cars, and they can use the existing lanes to reduce friction or solve their pain point, in this case avoid the logs. If you automatically assumed the answer was to increase capacity (lanes), you would be wasting resources, but this is what typically happens in enterprises because technology performance metrics are complex.

Another way to ease friction caused by inefficient processes is to balance the workload by changing the times that various processes run. These represent opportunities to make a positive impact by shifting processes to times when there are more available resources.

Today most people schedule the processes or jobs to run at zero, fifteen, thirty, or forty minutes after the hour. They don't look at the image above and say, "I can schedule this at 11:07 p.m. because there is not that much activity occurring then." Why not? Because either they don't have access to the data or if they do, they are unsure which performance metrics to look at.

Our systems are complex. We capture and process lots of data. Enterprises need to find an easier way to measure server health and efficiency and reduce friction at scale as the number of systems and data continue to grow. In our quest to improve the health of all technology services, we need to simplify how we see the data, measure the impact of change on the system, and interpret the next action to take at scale.

GOING BEYOND WHAT THE EYE CAN SEE

Visual dashboards leave it up to an individual, expert or not, to interpret and correlate many performance metrics to determine the health and performance of a server system and determine any actions to take. Accordingly, there can be significant variability in these decisions, depending on who is making them. There is no official scoring system for determining workload health. Proper measurement can be as complicated as the nature of the data itself.

Moreover, most of these visual dashboards and analyses related to server systems do not consider the health of the applications running on server machines; instead, they only focus on the server resources. Indeed, in cases where an application is running on a virtual server or in the cloud on a PaaS architecture, where the servers are operated and provisioned by a third-party provider, these visual dashboards might be close to useless, as they are unable to give any accounting apart from the server machine itself.

As a result visual dashboards do not provide insight or actionable assistance when unhealthy applications on a server negatively affect that server's performance. We need a system and method that produces guidelines to simplify the measurement of the health, performance, and capacity of a server—as well as the applications running on that server—to identify potential problems, define improvements, and implement solutions to increase performance and capacity (i.e., improve workload health).

A BETTER WAY TO APPROACH WORKLOAD HEALTH

I've developed a patent-pending approach to measuring server performance and application health that's focused on creating algorithms that pull together and interpret the right data to show, in very simplistic terms, the health of technology systems. If I can measure the health of a server, then I prescribe actions, which is what's missing from existing data observability and application platform monitoring (APM) tools. These don't tell you what to do; they merely identify problem areas and show more data. We need to change the paradigm from showing more data to finding better ways to understand and use the data we have.

We need to change the paradigm from showing more data to finding better ways to understand and use the data we have.

Machine learning and artificial intelligence are integral to my approach. *Machine learning* algorithms run against data to help identify anomalies and trends as the number of servers and systems continue to increase in size and complexity. Machine learning is simply a way to scale learning from the data and try to define, interpret, and ultimately detect anomalies and forecast the future workload. The model is continually learning from the data sets that are fed into it. And then it will adjust its thinking and recommend different observations based on the existing workload.

Artificial intelligence comes into play to continually learn and adjust the algorithms so we can create and refine actions based on their impact on the systems. AI analyzes the impact or outcome of the actions that are implemented in the system. For example, if I change setting X, did it solve the issue as expected; and if so, what was the impact? Next, if solution Y is implemented to solve the same problem, what was the impact? AI will look at all the fixes over time, analyze their respective impacts, and then recommend the most optimal adjustments and continually learn from the data.

How does someone supporting one hundred, or even one thousand, servers understand when there's an atypical behavior that's an anomaly? Machine learning and AI can help to identify patterns that we, as humans, could not possibly pick up on our own.

MEASURE, ACT, ASSESS (RINSE, REPEAT)

The key to addressing the workload analytics gap and gleaning more insights and actions that create impact from the available data is to develop a scoring model. The challenges today stem from determining the health, efficiency, performance, and capacity of technology systems. Most technology professionals do not have a way to express these in simple numbers or the expertise to interpret the raw data. If this wasn't the case, we could look at the capacity and the workload health score along with the growth metrics to forecast the capacity requirements for tomorrow or the peak business season.

Let's not stop there. What if there was a way to forecast the future for capacity and scaling the system out years? Today scaling systems and determining capacity requirements is a very manual and time-consuming process. I can envision a day when we can predict the target capacity requirements just as a 401(k) calculator can predict your retirement savings.

We need to go beyond tools that show more data, information tips, alerts, and help text, which are merely adding more color to dashboard images. We need actions. *Don't just tell me I have memory pressure within SQL Server. Tell me what setting to change in the system to alleviate the pressure and what the impact will be.*

How can we determine annual data growth or transactional growth? What applications are generating the most data in the organization? Even if you can make the right changes, how can you show that the system is healthier? The current tools don't tell you this. And to make things even more complex, we don't all talk the same language. Some analysts or administrators look at processor time percentage, while some look at memory; it depends on their skill set and background. What's more, there is a lot of information available on the internet and many "best practice" thresholds for different values, so people will bring their own knowledge bases to the table when making these assessments. We need to move away from making one-off judgments and turn interpretations into guidelines that can be followed and documented and then revised and followed some more across the industry.

In accounting you can run a profit and loss statement or balance sheet anytime for your business and see how the company is doing based on journal entries for money that came in and went out. These are considered "generally accepted accounting principles" that are standard across all organizations. Why do we not have any generally accepted principles of measurement across all technology systems?

Further, an accountant can look at an individual business line within a large corporation, but this input alone would be fairly meaningless if it wasn't combined with information from all of the other business lines. What if there was a way to consider all "lines" for technology systems together and overlay financial data across technology data to account for the costs of each company's technology environment? Or to determine the financial impact of code or processes before they are deployed to production and the business gets a bill for several thousand dollars that it wasn't expecting?

The challenge in technology today is there are no standards for capturing and measuring data and therefore no financial transparency. We need a better framework for bringing financial transparency to technology data, one that becomes universal within the industry. We need to capture not just data for data's sake but also the variables in data that impact the system, such as capacity, licensing, storage, memory, processing time, etc. With this approach, we'll be able to see where we're at today and, perhaps more importantly, where we'll be tomorrow. And we want to map the variables that contribute toward a healthier system based on whatever measurements, or indicators, we choose. We need to measure, determine action, assess impact, and then rinse and repeat.

Why this isn't already being done is anyone's guess. But we all provide assessments to clients and recommend actions, so how far of a leap could it be to automate this process in real time? Let's take some risks and make some assumptions so we can get on with the business of helping people improve their systems.

Back to the fitness app discussion. Why do you want to have a fitness check at all? You want to assess your environment and identify which factors contribute to your goals and which detract. In technology we need to produce a similar assessment that looks at the contributing and detracting factors to server health, identify the specific actions that need to change within your environment, and then map this analysis back to performance in a virtuous cycle.

Our goal is to understand where and how to optimize the system (i.e., if I make a configuration change or adjust this piece of code, does it make the server healthier or unhealthier?). Layer on top of this the cost associated with making the change versus the value it brings the company. *How does this change impact the whole ecosystem?* This is what I consider a "top-down" view, considering the impact on the entire organization.

If you wanted to identify the most impactful pieces of code to tune, how would you define *impact*? If you're the CFO, it would probably be those that are the costliest. But CFOs never look at this data compared to the data teams that are always reviewing it. Yet for some companies, financial impact may be the most important metric of all.

SCORING WORKLOAD HEALTH

When a company hires Fortified to perform an assessment of their technology environment, they want to know how healthy their technology is and where they can make improvements. Today we install WISdom, our proprietary workload intelligence tool, and are able produce a report with short- and long-term recommendations to increase the performance, efficiency, and health of the system. Once our clients start making changes, they come to rely on this tool, enabling them to see the impact of the changes on their systems. Through WISdom's KPIs on performance, efficiency, capacity, and resource health, they can easily see progress, measure success, and plan next actions.

Some clients take this a step further and have us rerun the assessment quarterly so they can have the next list of actions and continually improve their systems. Our clients can run our tool for themselves, watching their health scores go up, just as your fitness or sleep app score will go up if you're exercising more or sleeping better. We're doing the same thing for data; we're creating a *workload health score.*

> *If you sleep better, your exercise readiness score is going to go up. Wouldn't it be nice to know how ready your technology system is for the day ahead?*
>
> **—BEN DEBOW**

We're able to look at various workload health indicators across the system to see what's trending up and down (which systems are healthy versus unhealthy). We consider indicators from a database perspective, a server perspective, and an overall score. And everything that's going on in a system will be keyed off a workload health score, which will indicate when to schedule processes because it can determine when the system is busy and when there is no friction. With a deeper understanding of the data, we can be smarter about when we schedule processes. We can solve friction. With the right technology, customers can have the ability to schedule processes just as we do with Microsoft Outlook when we schedule meetings. "System, when are you free? OK, I can schedule something here."

We shouldn't have to wait for a consulting company to come in and provide this information. It should be available in real time, as the current tools don't tell us what to do *right now*. We need to know what it takes to make the server healthier. This can only be accomplished

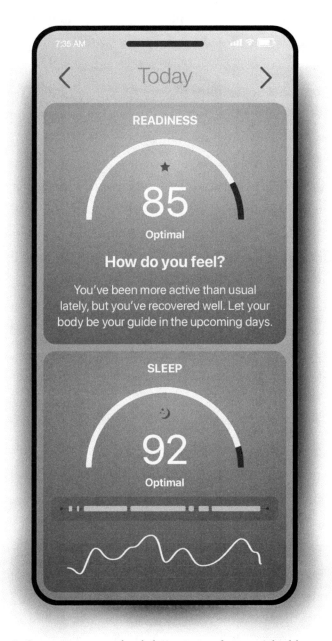

Figure 5: Our measures provide a holistic picture of a person's health using over twenty biometric signals to tell a person how ready they are for the day ahead.

through a new model that looks at the entire environment from the top down, rather than considering one server or one instance at a time.

Our approach provides enterprises with the ability to take resource planning to the next level for both infrastructure and people. If you know how many units of capacity you will need at any point in time or how many tasks must be done to keep the systems healthy, you can allocate resources accordingly.

Today most technology workers get their tasks from ticketing systems, which are populated from project requests, developer requests, and alerts. This is highly inefficient because it doesn't take into account everything that's required to manage, maintain, and optimize the systems. In my experience technology professionals only devote 10 to 20 percent of their time to focus on their tasks because they are struggling to keep up with the daily influx of problems and bugs that have become the accepted norm. This needs to change and will not change until we can all have a simple picture of health, capacity, and the work to be done to optimize the systems.

———————

Just as you want to see ten thousand steps pop up on your smartphone screen at the end of the day, you'll want to understand your optimal workload health score. If your score is too high or too low, you'll want to take the steps (pardon the pun) necessary to get it where it should be. And you'll want to identify and understand what those steps are so that they can make a difference to workload health today and for the future.

In the next chapter, we'll take a deeper dive into defining the health of a system considering the enterprise level and the system level and within specific technology platforms that run the business by considering what questions to ask depending on your role within the ecosystem.

KEY TAKEAWAYS

- Data for data's sake is virtually meaningless. It needs to be measured against goals to understand and optimize the health of systems.

- The use of server performance metrics often adds a layer of complexity that prevents correct, consistent, and efficient analyses— particularly when less experienced subject matter experts are overseeing it.

- Not having enough capacity is an obvious red flag. However, employing a *better safe than sorry* approach through oversizing resources often results in substantially higher costs and is ineffectual in preventing health and performance issues.

- Technology professionals need a new approach and method that produces metrics to simplify the measurement of the capacity and health of a server and the applications that run on it.

- Scoring workload health is an approach that can help identify potential problems, improvements, and solutions to increase performance and capacity and rightsize costs for the operation overall.

Defining Technology Health

One technology that has necessitated the need to scale workloads quickly is the quick response, or QR, code. Even the name itself implies speed and scale. Originally developed in the 1990s as an alternative to standard UPC barcodes used for production processes, such as tracking and item identification for the auto and food industries, QR codes found favor as a marketing tool in the early 2000s, with the advent of smartphones making it easy to scan the quirky images for a whole host of uses including issuing coupons and promotions.

Fast-forward to the COVID-19 era, when people were (and still may be) afraid to touch menus for fear of germs. Instead, restaurants found they could save on paper menus by enabling easy access to a QR code leading to their flatbread pizzas, sliders, and brussels sprouts. Wary travelers don't have to touch an airport kiosk to print their plane tickets if they have a QR code emailed to them. All you need is a smartphone and—poof!—instant menu, instant boarding pass.

Shameless plug: Use this QR code to visit my website, www.bendebow.com!

One so-called success story of QR codes for marketing involves the cryptocurrency company Coinbase, which came up with an arguably brilliant idea for their $15 million sixty-second commercial in the 2022 Super Bowl. It was a QR code. That's it—for sixty seconds, with some bouncy music playing as the code bounced along on the screen. From a marketing perspective, it was a success—twenty million people came to Coinbase's site in just one minute after scanning the QR code on their smartphones.[8] Coinbase's marketers were thrilled as the app went from 186th place to 2nd place in Apple's App Store.[9] The technology staff? They didn't have as much to celebrate.

Both the Coinbase site and application crashed due to all of that unplanned traffic. The outage only lasted a few minutes, but the damage was enough to send the share price of Coinbase stock down 5 percent.[10] While all ended well for Coinbase—the problems were fixed in less than an hour, and the added publicity from the whole mess more than likely made up for any losses they incurred—other companies simply cannot afford site outages like this.

In addition to reputation damage, outages cost! Gremlin has a tracker on their website (https://gremlin.com/ecommerce-cost-of-downtime) that measures the cost of downtime for more than twenty major retailers.[11] Leading the pack by a wide margin is, you guessed it, Amazon—a single minute of downtime for Amazon costs $220,000, or $13.2 million per hour.

---

8   Sarah Roache, "Coinbase's Super Bowl Ad Brought 20 Million People to Its Site in a Single Minute," Protocol, February 14, 2022, https://www.protocol.com/bulletins/coinbase-super-bowl-results.

9   Ibid.

10   William Suberg, "Coinbase Stock Falls 5% Pre–Wall Street as Bitcoin Price Dip Adds to Super Bowl Woes," Cointelegraph, February 14, 2022, https://cointelegraph.com/news/coinbase-stock-falls-5-pre-wall-street-as-bitcoin-price-dip-adds-to-super-bowl-woes.

11   "The Cost of Downtime for the Top US Ecommerce Sites," Gremlin, accessed November 23, 2022, https://www.gremlin.com/ecommerce-cost-of-downtime/.

Blue Triangle further estimates that a five-minute outage during Cyber Monday can cost major retailers over $1 million in revenue, and even a one-second site slowdown can cost a business 6 to 13 percent of their total revenue.[12] Although Cyber Monday has been around since 2005, retailers from Office Depot to GameStop to Walmart still experience slowdowns and outages in the highly trafficked shopping days post-Thanksgiving.[13]

THE REAL IMPACT OF OUTAGES

It may be no surprise that when the behemoths of the cloud experience an outage, it has a widespread effect, but what may be surprising is how often these occur as well as the breadth of impact.

A "huge" outage occurred for Amazon Web Services (AWS) at a data center in Virginia on December 6, 2021, impacting businesses and services beyond Amazon's own Alexa and Ring doorbell across the world *for almost six hours* including:[14]

---

12   Adam Wood, "Cyber Monday 2021: The Fastest and Slowest Retail Websites," *Blue Triangle*, accessed November 22, 2022, https://blog.bluetriangle.com/cyber-monday-the-fastest-and-slowest-retail-websites.

13   Paul Conley, "Website Outages, Slowdowns Hit Dozens of Retailers during Cyber 5," Digital Commerce 360, November 29, 2021, https://www.digitalcommerce360.com/2021/11/29/website-outages-slowdowns-hit-dozens-of-retailers-during-cyber-5/.

14   S. Dent, "Amazon, Reddit, Twitter, and Twitch Impacted by Huge Network Outage," Engadget, June 8, 2021, https://www.engadget.com/a-huge-outage-is-affecting-large-swaths-of-the-internet-102354305.html; Ben Gilbert, "Why Everything from Netflix to Nintendo Goes Offline When Amazon's Servers Have Issues," Insider, December 11, 2021, https://www.businessinsider.com/why-does-everything-break-when-amazon-servers-go-down-2021-12; J. Condit, "Amazon Web Services Went Down and Took a Bunch of the Internet with It," Engadget, December 7, 2021, https://www.engadget.com/amazon-web-services-outage-dec-2021-173157290.html.

- Netflix
- Nintendo
- TindMcDonald's
- Sweetgreen
- Disney+
- Roku
- Associated Press
- The *New York Times*

- The *Guardian*
- *Vice*
- Twitch
- Reddit
- Hulu
- HBO Max
- Shopify
- Twitter

- Stack Overflow
- GitHub
- Gov.uk
- Quora
- PayPal
- Vimeo

Business Insider referred to the situation as "a total mess—the result of a surprisingly large group of major companies depending entirely on Amazon for base level functionality."[15] Think about how much revenue must have been impacted during a six-hour outage at any one of these companies, not to mention the global impact including thousands of other businesses worldwide.

Here's how AWS responded to the event:

We immediately disabled the scaling activities that triggered this event and will not resume them until we have deployed all remediations. Our systems are scaled adequately so that we do not need to resume these activities in the near-term. Our networking clients have well tested request back-off behaviors that are designed to allow our systems to recover from these sorts of congestion events, but, a latent issue prevented these clients from adequately backing off during this event. This code path has been in production for many years but the automated scaling activity triggered a previously unobserved behavior. We are developing a fix for this issue and expect to deploy this change over the next two weeks. We have also deployed additional network configuration that protects potentially impacted networking devices

---

15   Gilbert, "Why Everything from Netflix to Nintendo Goes Offline."

even in the face of a similar congestion event. These remediations give us confidence that we will not see a recurrence of this issue.[16]

Here is a list of other outages occurring for AWS over the past ten years.

YEAR	MONTH & DATE (AVAILABLE)	EVENT TYPE	DETAILS
2011	4/21	Outage	At 12:47 AM PDT on 4/21, an invalid traffic shift prior to network upgrade caused EBS instances to lose connectivity to one another with an availability zone of US-East-1 region. Once the errors were localized to just one availability zone, the EBS recovery These connectivity errors impacted EBS volume and EC2 instances in multiple availability zones and caused issues for customers until full recovery at 3:00 PM PDT on 4/24.
2011	8/7	Outage	Power lost in Ireland, EU West region, causing disruption and outage. "Service disruption began at 10:41 AM PDT on 8/7th" (also mentioned but distinct from the outage mentioned above; it happened around the same time as the US outage). Due to followup issues, full restoration of EBS and RDS took in the order of days.
2011	8/8	Outage	EC2 went down around 10:25 p.m. Eastern in Amazon's U.S. East Region. The cloud outage lasted roughly 30 minutes, but took down the Web sites and services of many major Amazon cloud customers, including Netflix, Reddit and Foursquare. The issue happened in the networks that connect the Availability Zones to the Internet and was primarily caused by a software bug in the router.
2012	6/29	Service disruption	A major disruption occurs to the EC2, EBS, and RDS services in a single availability zone (due to a large scale electrical storm which swept through the Northern Virginia area).
2012	10/22	Outage	A major outage occurs (due to latent memory leak bug in an operational data collection agent), affecting many sites such as Reddit, Foursquare, Pinterest, and others.
2012	12/24	Outage	AWS suffers an outage, causing websites such as Netflix instant video to be unavailable for customers in the Northeastern United States.

16   "Summary of the AWS Service Event in the Northern Virginia (US-East-1) Region," AWS.amazon.com/message/12721/.

YEAR	MONTH & DATE (AVAILABLE)	EVENT TYPE	DETAILS
2013	9/13	Outage	AWS US-East-1 region experienced network connectivity issues affecting instances in a single Availability Zone. We also experienced increased error rates and latencies for the EBS APIs and increased error rates for EBS-backed instance launches.
2014	11/26	Service disruption	Amazon CloudFront DNS server went down for two hours, starting at 7:15 p.m. EST. The DNS server was back up just after 9 p.m. Some websites and cloud services were knocked offline as the content delivery network failed to fulfill DNS requests during the outage. Nothing major, but worthy of this list because it involved the world's biggest and longest-running cloud.
2015	9/20	Outage	The Amazon DynamoDB service experiences an outage in an availability zone in the us-east-1 (North Virginia) region, due to a power outage and inadequate failover procedures. The outage, which occurs on a Sunday morning, lasts for about five hours (with some residual impact till Monday) and affects a number of related Amazon services include Simple Queue Service, EC2 autoscaling, Amazon CloudWatch, and the online AWS console. A number of customers are negatively affected, including Netflix, but Netflix is able to recover quickly because of its strong disaster recovery procedures.
2016	6/5	Outage	AWS Sydney experiences an outage for several hours as a result of severe thunderstorms in the region causing a power outage to the data centers.
2017	2/28	Outage	Amazon experiences an outage of S3 in us-east-1. There are also related outages for other services in us-east-1 including CloudFormation, autoscaling, Elastic MapReduce, Simple Email Service, and Simple Workflow Service. A number of websites and services using S3, such as Medium, Slack, Imgur and Trello, are affected. AWS's own status dashboard initially fails to reflect the change properly due to a dependency on S3. On 3/2, AWS reveals that the outage was caused by an incorrect parameter passed in by an authorized employee while running an established playbook, that ended up deleting more instances than the employee intended.

YEAR	MONTH & DATE (AVAILABLE)	EVENT TYPE	DETAILS
2018	3/2	Service degradation	Starting 6:25 AM PST, Direct Connect experienced connectivity issues related to a power outage issue in their US-East-1 Region. This caused customers to have service interruptions in reaching their EC2 instances. Issue was resolved fully by 10:26 AM PST.
2018	5/31	Outage	Beginning at 2:52 PM PDT a small percentage of EC2 servers lost power in a single Availability Zone in the US-EAST-1 Region. This resulted in some impaired EC2 instances and degraded performance for some EBS volumes in the affected Availability Zone. Power was restored at 3:22 PM PDT.
2019	8/23	Outage	A number of EC2 servers in the Tokyo region shut down due to overheating at 12:36pm local time, due to a failure in the datacenter control and cooling system.
2019	8/31	Outage and data loss	The US-EAST-1 data center suffered a power failure at 4:33am local time, and the backup generators failed at 6am. According to AWS, this affected 7.5 percent of the EC2 instances in one of the ten data centers in one of the six Availability Zones in US-EAST-1. However, after restoring power, a number of EBS volumes, which store the filesystems of the EC2 cloud servers, were permanently unrecoverable. This caused downtime for companies such as Reddit.
2019	10/22-23	Service degradation from DDoS	AWS sustained a distributed denial of service attack which caused intermittent DNS resolution errors (for their Route 53 DNS service) from 10:30am PST to 6:30pm PST.
2020	11/25	Outage	Beginning at 9:52 AM PST the Kinesis Data Streams API became impaired in the US-EAST-1 Region. This prevented customers from reading or writing data.
2021	12/7	Outage	Beginning at 10:45 AM PST "an impairment of several network devices" in the US-EAST-1 Region caused widespread errors in all AWS services. The root cause has been mitigated by 4:35 PM PST, but service recovery was still underway causing localized ongoing impairment.
2021	12/15	Outage	Region us-west-1 was unavailable for about 30 minutes.

YEAR	MONTH & DATE (AVAILABLE)	EVENT TYPE	DETAILS
2021	12/22	Outage and potential data loss	Power loss in us-east-1 for about 1 hour, followed by extended recovery procedures. AWS attributed the failure to a single availability zone, USE1-AZ4.

Figure 6: Amazon web outages over the last ten years, as reported by Wikipedia, accessed on November 23, 2022, https://en.wikipedia.org/wiki/ Timeline_of_Amazon_Web_Services#Amazon_Web_Services_outages.

Does this sound familiar?

The CEO of an accounting software company wants to know if their solution is ready for peak season between January and April, especially for one day in particular—April 15. So she asks her CIO, who then asks a director or VP, and he, in turn, asks the managers of the application team, the infrastructure team, and the data team. These people then ask the subject matter experts on their respective teams, who proceed to look at their own specific servers that run the tax application. They all pull some data from their SME tools and identify areas of concern through busy, color-coded dashboards. They must infer what the capacity is. And then everyone runs that information up the chain of command in what I call a "bottom-up" approach.

Adding to this complexity, a lot can change for tax applications in one year based on new accounting rules. Based on my experience, one additional year of data can deviate between 20 and 100 percent or more from the previous year, depending on the organization. On top of that, companies deploy many new features and add more transactions and users, which changes workload and capacity requirements.

So the truth is this business may have no real idea what the health of their system is if it's running fine and there aren't any flags. If you're an accounting firm trying to get through the new accounting rules for 2022, you likely filed an atypical number of extensions because your systems weren't prepared to accommodate the changes in workload, not to mention the manual hours required from your staff. And if you're anything like Coinbase and you get ten or twelve times the load coming in on one day and the system is only used to running a certain amount of data, it's very hard to calculate, especially if any new features are being supplied in addition to the increased traffic.

SMEs may be able to spot areas of potential concern, but it would be a physical impossibility for them to foresee every potential issue on every app living on every server in the ecosystem. It would take a lot more time, analysis, and testing to understand whether there is enough capacity or if the system will perform fast enough to support all the "right" actions that need to occur at the same time. Performance testing alone, as it's typically been done, is not enough when you have no way of gauging what the actual downside might be along with other potential unknown variables.

So what could Coinbase and these major retailers have done differently to define their workload health, resiliency, and capacity? How could they have known if their system was healthy enough to withstand the traffic jams that brought them down?

For starters I maintain we need to start at the top, not the bottom. Executives need to ask different questions of the technical staff and business owners of the applications:

We need to start at the top, not the bottom.

- What are the goals for the peak season across all applications?
- What is the expected number of users and transactions, by application, at the peak time?

- What are the performance SLAs (service-level agreements) that need to be met (e.g., the average request must be resolved in two seconds)?
- What is the estimated capacity for each server and application within the server?
- What are areas of opportunity to optimize the application, and what is the expected impact?
- For each change, does this increase performance by 1 percent or 5 percent? (Most people cannot anticipate the impact of changes to the application they make. They wait until the gauge on the monitoring chart changes to understand the impact in real time.)

The other important questions that are missing from the discussion today pertain to measurement:

- What are the KPIs we will use to validate that our applications are healthy and optimized?
- How do these application KPIs get rolled up to enterprise KPIs across the enterprise?

I've seen this situation of relying on the old way of doing things ring true in many companies, even ones that have a lot of experience running complex systems with an array of monitoring tools, performance reports, and dashboards in place. Then the business is inevitably left to wonder, *Why is there a gap in performance, and why are we not able to answer the basic questions above?*

At the time of this writing, Fortified is actively engaged in about ten scaling and capacity planning projects for our clients, which says something about today's tools. If it was that easy to use the tools, our clients wouldn't require our services. So what do we need to change about how we interpret data and use tools to give us clear answers on capacity, efficiency, and workload health from the enterprise down to the application services?

The only way we can really know definitively today if the system is equipped to handle what may come tomorrow is to create a duplicate workload in a parallel environment and conduct testing to identify what will make the system more efficient or strong enough to be able to handle different workloads. This is not to say that all applications need to have a parallel environment to load test, as this is usually reserved for the largest applications within each enterprise. Since there are different complexities in play and a significant drain on resources to accomplish this, we need a more simplified, straightforward way to understand the health and capacity of a system. The other 99 percent of enterprise applications and services need clear KPIs and ways to measure success—before April 15, Super Bowl Sunday, or whatever do-or-die parameter makes sense for the organization.

This, in my opinion, is ground zero for simplifying this story of understanding the health, efficiency, and capacity of the system at any time, not of any one server or application running on that server but of the whole system, from the top down. What if you have one thousand servers in your application, such as a company like Intuit does? And that's just for QuickBooks alone—one app among hundreds of applications. You would have to recreate the process hundreds of times and then aggregate all those results to be able to make a true assessment.

So what would be a better way to measure the health of your entire system at any given time? Before we set out to answer this question, let's consider the basic levels of performance within the technology enterprise.

ORGANIZATIONAL STRUCTURE FOR ENTERPRISE TECHNOLOGY

Each level of an organization has a stake in the health of the technology environment. Here we look at how each level is connected to the environment, who within the corporate hierarchy plays a role, and how they approach defining server health from their own unique vantage point.

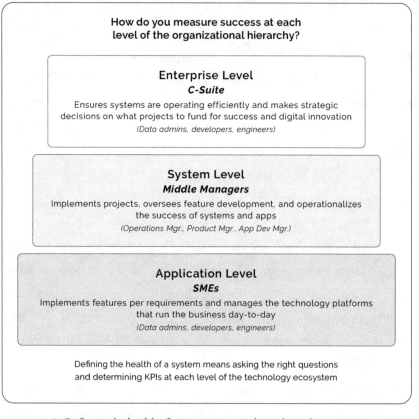

How do you measure success at each level of the organizational hierarchy?

Enterprise Level
C-Suite
Ensures systems are operating efficiently and makes strategic decisions on what projects to fund for success and digital innovation
(Data admins, developers, engineers)

System Level
Middle Managers
Implements projects, oversees feature development, and operationalizes the success of systems and apps
(Operations Mgr., Product Mgr., App Dev Mgr.)

Application Level
SMEs
Implements features per requirements and manages the technology platforms that run the business day-to-day
(Data admins, developers, engineers)

Defining the health of a system means asking the right questions and determining KPIs at each level of the technology ecosystem

7: Defining the health of a server means asking the right questions and determining KPIs at each level of the technology ecosystem.

APPLICATION LEVEL

Application engineers, infrastructure engineers, and database administrators (DBA) make up the front-line support teams for either building the applications or supporting the infrastructure and data that the applications run on. The engineers and DBAs are responsible for making sure the applications run correctly, the infrastructure is upgraded as needed, and the systems have the right amount of capacity and resources to meet the business objectives. They are often called subject matter experts because each is an expert or focused on supporting their own technology or application.

SMEs determine the health of their applications by looking at tools, such as LogicMonitor, SolarWinds, or the native monitoring tools that Oracle and Microsoft provide. While these tools distill information into various graphs and dashboards, the majority stop short of indicating what to change to fix the issue or improve the system. Therefore, today's SMEs need to be even more skilled to operate complex technology systems and applications, as they can't rely on the tools alone. Long term, unless something changes, we will have an ever-widening resource gap as the data and number of technology services continue to grow, while the number of true specialists fail to keep pace.

SYSTEM LEVEL

Managers at the system level do not have as cut-and-dried a method as SMEs for determining technology health, as they have to rely on the interpretations of their teams. These managers are primarily interested in three key indicators: availability, stability, and performance. If the system is up and no one is complaining of slowness or other issues, they may feel the system is healthy from their perspective because it's doing what it's supposed to.

It's my contention that those responsible for applications need to go beyond the status quo to understand the following:

- How much more efficient *could* the application and/or code be?
- Where are opportunities to optimize the application or reduce cost?
- Can the system scale and process the business workload expected for peak season this year?
- What will my capacity requirements for the system be over the next one to three years?

Most organizations focus on meeting the functional specifications when building an application, which is important. But they never ask these questions until there is an issue, an assessment is performed, or cost cutting becomes mandated. And then they are acting in a reactive rather than proactive way, addressing problems, waste, and inefficiencies that could potentially have been avoided.

ENTERPRISE LEVEL

Executives at the enterprise level, such as CIOs, CDOs, and CTOs, are tasked with digital transformation and leveraging technology to enable the business. The questions commonly asked from this view are as follows:

- How will data and technology be used to transform the business?
- How healthy are my systems today (depending on their own definition of *health*)?
- Are our mission-critical applications performing well?
- Are our clients happy? Did we process all our orders successfully and in a timely manner? Did we miss any SLAs?

CFOs also operate at the enterprise level but have a different view centered on the bottom line. So add these to this list:

- Are we on budget for hosting and supporting all the technology services for the organization?
- Why are the cloud costs higher than we budgeted for?
- What changes can be made to reduce the cloud costs?

Without the right, timely information reflecting the health, capacity, and costs of their technology systems, how can executives at the enterprise level drive results that will impact the very future of the business overall?

It's important to examine the overall structure of an organization to understand the different vantage points on server health. Each role within each level of the technology ecosystem has different ways to measure success. This existing hierarchy serves to keep us from identifying a unified set of KPIs that will definitively tell us the health of our systems. In the absence of a scoring model or universal benchmarks to consult, there are four major indicators that will influence success at each level of the hierarchy and for each role within it.

KPIS TO GAUGE THE HEALTH OF TECHNOLOGY

While there is no standardized way to measure the health of technology systems, most technology professionals map their KPIs to a commonly understood set of measurements. While each organization makes its own determination of the specific KPIs to map against each aspect of technology performance, there are four basic determinants that are generally regarded to measure the health of an organization's technology environment: availability, stability, performance, and capacity.

Here I'll present a high-level view of how these are currently applied to technology performance measurement, with the understanding that many of you reading this may have a slightly different approach based on your organization's purpose and system's function.

1. AVAILABILITY

System availability is arguably the easiest to understand. For what percentage of time is the system experiencing uptime, meaning the system is up and available? A KPI is set as a percentage of this time for a given period—a day, month, year. For some systems 95 percent uptime may be acceptable; for others it must be 99.995 percent. It depends on what the system is being used for and what the downtime represents.

Tools, such as an SLA calculator available from Uptime.is, make it very easy to estimate the amount of downtime that's acceptable within an SLA. For example, an organization that expects an uptime, or availability, of 99.99 percent, per their SLA, can accept 0.01 percent of downtime, which translates into fifty-two minutes and thirty-two seconds per year. That may sound reasonable, but if the organization is a hospital ER, this nearly one hour of downtime could potentially cost lives.[17]

2. STABILITY

The stability of a system presents a few more challenges to quantify. *Merriam-Webster* defines *stability* as "the strength to stand or endure."[18] That makes sense for systems too, the ability to endure disruption and crisis. How likely is your system to continue running in the long term when it is forced beyond normal capacity? In other words, how *reliable* is it? What is the point at which the system breaks down or fails to function? Or how *predictable* are its processes? Does this process perform the same way every day, or does it sometimes run longer or use more resources? In short, reliability and predictability are drivers

---

17   Uptime.is, accessed November 28, 2022, https://uptime.is/.

18   *Merriam-Webster*, s.v. "stability," accessed November 23, 2022, https://www.merriam-webster.com/dictionary/stability.

of stability. While the system may be currently delivering based on a certain set of variables, can it be expected to continue to perform if one or more of those conditions change?

Essentially stability refers to how strong your system is now, and reliability and predictability have to do with the extent to which it can be expected to remain strong tomorrow. The only way to truly measure the strength of a system is to run tests in a parallel environment that mimic what could happen in a real-world situation. Mapping KPIs to stability, including reliability and predictability, is not as simple as availability because there is no way to account for every possible scenario that could impact a system. And how many resources would it take (people, time, money) to implement these test scenarios if we tried?

Consider this scenario regarding predictability. Each year retail businesses try to predict how much inventory they will need and how much sales volume they will have during the holiday season. And every year retailers think they have it figured out, but a new factor comes into play that turns all those predictions on their head. For 2020 it was the COVID-19 pandemic; for 2021 it was the supply chain crisis. Predicting anything is hard, yet people insist on trying to predict everything from the behavior of the stock market to the impact of climate change to who will become the next president of the United States.

Technology systems are no different. Organizations need and expect their applications, infrastructure, and services to provide stable, reliable, and predictable performance over time.

3. PERFORMANCE

Your system may be available, it may be stable, and it may even be reliable, but is it *performing* optimally? Or are there performance anomalies that are impacting the customer experience? Just because a system is functioning at 100 percent uptime doesn't mean its level of performance is acceptable or couldn't be improved.

At Fortified one indicator of performance might be, *Are we meeting what we promised the client in their service-level agreement?* If the percentage of requests outlined was met within a certain time frame, then we've achieved our performance goals. But who sets the goals, and how do they vary among organizations? SLAs can vary greatly between organizations because it is difficult to put a number on the line when so many complexities are involved. Further, there may be a different set of owners involved in the task, each with a different set of expectations of what should be delivered.

In short a smaller organization may have very different performance needs than a large enterprise. Serena Williams may commit to a different level of performance than I do, but both of our goals are just as valid.

The other key question regarding performance that doesn't get as much attention, I believe, is efficiency—not just "are we meeting the goal?" but also "how *efficiently* is the system running in order to meet it?" In other words how much waste is there in the system, which, of course, translates into dollars. The code may meet performance goals, but how efficient is the process? Can the code be optimized more and use fewer resources, which would reduce the costs to execute, increase capacity of the system, and most likely increase the performance? How do we capture this opportunity to save resources and money?

Just as we map KPIs to each technology outcome, we must also define and measure the efficiency of the system overall. How can we reasonably be expected to implement any changes at scale across an enterprise with six hundred applications and ten-thousand-plus servers if our system is inefficient? SLA performance goals may be determined, but we need to define and measure them in a way that is digestible across the enterprise. We need to develop a standard set of formulas and technologies that can be applied to easily measure, manage, and report on deliverables across systems and applications and ultimately across organizations.

How can you quantify an acceptable level of performance and efficiency for a technology system? By looking at the levels of traffic jams that are causing the system friction and by understanding where the system is experiencing overcapacity or undercapacity because not having enough capacity can result in poor performance, while too much can result in wasted resources.

4. CAPACITY

It's hard to talk about any KPIs without discussing capacity. Capacity basically means, is there enough? Are there enough resources to run the system so that it can do what it's supposed to do? When people think in terms of capacity today, they don't think in terms of measurement. There is no single way to vocalize what the true capacity of a system is at any point in time. Instead, capacity is usually referenced in terms of rightsizing systems. Do we have enough? But what defines *enough*? The system may be experiencing overcapacity, but by how much? There's no number for that, no standard formula to measure across systems and applications. But could there be one? If we represent capacity at a very high level, we can look at memory and CPU or storage. How many areas within each system have undercapacity, how many have overcapacity, and how many are functioning optimally?

Capacity today is often calculated at the host or server level. These computations don't look at the internal health of the applications running on the server and therefore are missing a key part of the story. This omission increases the risk of negative impact on the application's performance and misrepresents the true level of capacity at the enterprise level.

Capacity is also more dynamic than these conventional metrics assume. For example, capacity cannot be measured just by looking at processor time percentage as most people do today. Consider the cloud and virtualization. The system administrator can easily increase/decrease server resources in minutes to rightsize the system and optimize resources.

But again this only considers the server level while not considering application health. Why aren't administrators looking at the health of the applications while adjusting server resources? Because it is challenging, and a common gap is that the infrastructure team typically does not know code and the applications' health metrics. Still, it should not be overlooked.

Most servers aren't processing the same workload throughout any twenty-four-hour period, so technically speaking, there is always excess capacity during the day. How can we represent this capacity to SMEs so they can schedule processes at different times or optimize code during those times? How do we aggregate the capacity data up to the enterprise level so that managers and executives can plan for system anomalies and perform enterprise capacity planning?

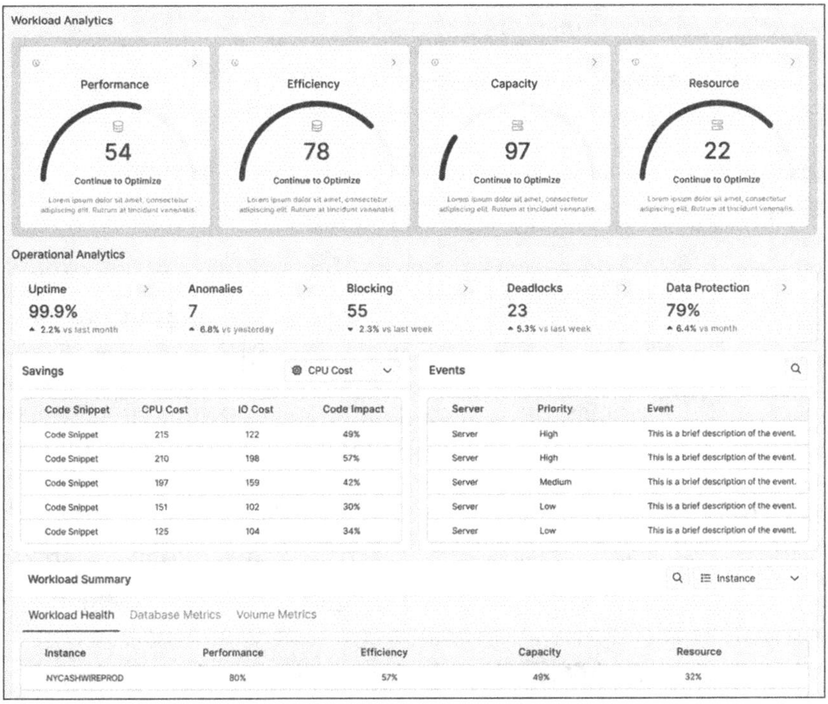

Figure 8: Fortified workload analytics dashboard—a better way to view database activity and gauge the impact of code changes on a system.

What we're really asking is, How do today's algorithms need to change to truly represent server capacity while also taking into account the performance and health of applications?

THE NEED FOR STANDARDIZATION

While availability may be commonly understood and measured, I believe the technology industry is lacking a standard definition for *stability* (including *reliability* and *predictability*), *performance* (including *efficiency*), and *capacity*. Why is this important? Because without a standardized vernacular, we have no way to map KPIs and truly measure the health of a system. And we have no way to know if we are improving things when we make changes.

When we define *server health*, we want to define it very subjectively, according to the goals of a particular organization, and very easily at any point in time as measured against the four conditions. We need a way to instrument and calculate these KPIs that will not only tell us how we are doing at a certain time but also make us understand the impact of system changes on the KPIs so we can measure progress.

It's not my intent with this book to standardize these definitions or create the formulas that will enable their measurement universally.

> **Reject the status quo and start to mature and standardize technology KPIs that enable all stakeholders to measure success the same way.**

But as I said in this book's introduction, we need a collective call to action to reject the status quo and start to mature and standardize technology KPIs that enable all stakeholders to measure success the same way. My eventual goal is to bring awareness of these shortcomings to the surface—as the old saying goes, "If you can't measure it, you can't manage it."

Let's come together and create better ways to measure success for technology professionals. If we can start to agree on not just what the KPIs are but also what they mean, then we can start to figure out new ways of measuring server health and performance. If we could accomplish this by just a small percentage, think of the gains that could be made for businesses, for industries, even for the economy overall. For the fundamental question still remains—how healthy is your technology environment?

In the next chapter, we'll consider the real reason we care about any of this as we tackle the universal goal of improving business productivity—and the one thing you must have in order to achieve it.

KEY TAKEAWAYS

- Defining the health of a system means asking the right questions at the application level (SMEs, data admins), system level (middle managers), and enterprise level (CIOs, CTOs, CFOs).

- The current methods and tools do not provide a true picture of server health because they only consider the health of a server while overlooking the health of the applications that are running on the server.

- To truly gauge the health of a server, it would require creating a parallel system that can be continually tested to predict the performance and capacity at both the server and application levels. This is generally not feasible due to limited resources and time.

- KPIs must be mapped to the levels of availability, stability (reliability and predictability), performance (efficiency), and capacity of systems in order to provide a true gauge of server health.

- We need a collective call to action to reject the status quo and be willing to ask a new set of questions that probe further to better define and understand the health of our systems from a more holistic, top-down perspective.

CHAPTER 3

The Most Important KPI You're Not Thinking About

B ack in the day when mainframes roamed the earth and each computer had 256 MB of memory, application developers had no choice but to care about resource consumption because they had a limited amount. When the developers would compile their code, they made sure it ran with a certain level of CPU and memory, or else it would not run at all.

Compare this to today, when a developer writes code as quickly as possible because they have twelve features to create in the next application sprint. If they have any issues with the code executing quickly enough once the system is promoted to production, typically the administrators add more CPU or memory to make it run faster. This story repeats itself thousands of times a day around the world for every slow-performing application. Why? Because it's a lot easier to add hardware than it is to fix code.

A popular tenet of the tech industry is "If it's not broken, don't fix it." In my experience, if it's not causing an outage, I don't know of a single application I've worked on in which the team has ever gone back and fixed something to make it more efficient. Why? Because they're too busy putting out fires or developing new features. No one is focused on making his or her application better.

I talked a lot about capacity and performance in the previous chapter. Why is any of this important? Why do we care about how much capacity there is in our systems or how they are performing? Ultimately it comes down to one thing that influences and impacts just about everything else we're talking about in this book—efficiency.

Every system strives to be more efficient, whether that's a family, school, city, business, corporation, or society. Inefficiency breeds waste, and no one likes waste. What if an application costs one thousand dollars to run but only needs to cost one hundred dollars to get the same results?

To increase efficiency, we must reduce the resources each application process or code uses, thereby lessening the burden on the server upon which everything is running. If you're not using the capacity, you don't need it, at least for now. However, we need to shift focus to the bigger, long-term capacity plays at hand if we want to make a real impact on efficiency. And we need to define a measurement standard for efficiency across the industry and over time.

FUEL EFFICIENCY FOR APPLICATIONS

You wouldn't buy a car without considering how many miles per gallon it gets. Why? Because cars have different engines, body types, engineering, and parts, all of which lead to a standardized fuel efficiency rating. So how can you go about assessing these important criteria before you make a purchase?

You can ask the dealer because, well, we all know we can trust anything a car dealer says. You can ask other people who've owned the same make and model. Or you can search for "miles per gallon," or MPG, figures online. One site, WhatCar.com, offers a measure called "true MPG" derived from "laboratory tests [that] use real-world driving routes to show what fuel economy you can really expect from your car."[19] They also assert, "True MPG is not only fully reflective of real-world performance but much more realistic than the official government fuel economy figures that car manufacturers have to quote."

There is a standard that exists that says one gallon of gas yields a certain number of miles, and this comparison holds true against any type of car built in any year by any manufacturer. Similarly, we need a "true MPG" measure for efficiency in technology.

By having a standard measurement of efficiency, each application and piece of code in the application can be assessed during the development life cycle. Similarly we can measure existing applications in production to see which applications or servers have more efficient workloads and which ones need to be optimized.

Another important consideration these days is the cloud. If an application is not efficient, do we want to migrate it to the cloud since we know it will cost more?

ARE WE RUNNING OUT OF RESOURCES IN THE CLOUD?

Many think the cloud capacity is endless, but it is not. Microsoft, Amazon, and others have literally run out of resources in certain regions or data centers. Cloud-based services, such as Microsoft Azure, use terms like *massive* and *enormous* to describe the storage options available with their platform, as not only businesses but also entire US cities are trusting them with their information.[20] And these words may

---

19   "True MPG Calculator," What Car?, accessed November 23, 2022, https://www.whatcar.com/truempg/mpg-calculator.

20   Microsoft Azure, accessed November 23, 2022, https://azure.microsoft.com/en-us/free/storage/search/.

not be enough to describe the magnitude of the situation. Microsoft plans to build fifty to one hundred new data centers a year to keep up with demand for its cloud services as global internet traffic is expected to double by 2022.[21]

If we wanted to put the whole city
in the cloud, we needed Azure.

—CHRIS MCMASTERS,
CHIEF INFORMATION OFFICER, CITY OF CORONA, CA

In 2019 the World Economic Forum reported that, in the coming years, common units of measure, such as megabytes, gigabytes, and even terabytes, will begin to seem quaint—because the entire digital universe is expected to reach forty-four *zettabytes* by 2020. If this number is correct, it will mean there are forty times more bytes than there are stars in the observable universe.[22]

---

21   Justine Calma, "Microsoft Ramps up Plans to Make Its Data Centers Less Thirsty," Verge, October 27, 2021, https://www.theverge.com/2021/10/27/22747394/microsoft-data-centers-water-drought-climate-change-energy-emissions.

22   Jeff Desjardines, "How Much Data Is Generated Each Day?," World Economic Forum, April 17, 2019, https://www.weforum.org/agenda/2019/04/how-much-data-is-generated-each-day-cf4bddf29f/.

DEMYSTIFIYING DATA UNITS

From the more familiar 'bit' or 'megabyte', larger units of measurement are more frequently being used to explain the masses of data

Unit	Value	Size
b bit	0 or 1	1/8 of a byte
B byte	8 bits	1 byte
KB kilobyte	1,000 bytes	1,000 bytes
MB megabyte	$1,000^2$ bytes	1,000,000 bytes
GB gigabyte	$1,000^3$ bytes	1,000,000,000 bytes
TB terabyte	$1,000^4$ bytes	1,000,000,000,000 bytes
PB petabyte	$1,000^5$ bytes	1,000,000,000,000,000 bytes
EB exabyte	$1,000^6$ bytes	1,000,000,000,000,000,000 bytes
ZB zettabyte	$1,000^7$ bytes	1,000,000,000,000,000,000,000 bytes
YB yottabyte	$1,000^8$ bytes	1,000,000,000,000,000,000,000,000 bytes

A lowercase "b" is used as an abbreviation for bits, while an uppercase "B" represents bytes.

Figure 9: By the byte.

How much data do we create *in one day*? Raconteur created an infographic to illustrate the exponential growth of data fueled by the internet of things and use of connected devices—in the context of one day—which they describe as "hard to comprehend." Some interesting stats they uncovered are as follows:[23]

- 463 exabytes of data will be created every day by 2025.
- 294 billion emails are sent every day.

23 Raconteur infographic citing IDC, Radicati Group, Statista, Intel, Facebook, and Twitter as sources, accessed November 28, 2022, https://www.raconteur.net/infographics/a-day-in-data/.

- 4 petabytes of data are created by Facebook every day.
- 28 petabytes of data are generated by wearable devices every day.
- 4 terabytes of data are used by a connected car every day.
- 500 million tweets are sent every day. (Remember, it was a big deal when Ashton Kutcher reached one million Twitter followers? Today he has seventeen million.[24])

While experts may disagree as to whether the supply of storage is infinite, the real point, I believe, is best summed up by Christian Fuchs, professor of media and communication studies at the University of Westminster, who said in this 2021 article for Gizmodo:[25]

Do we want endless storage of almost all aspects of our lives? What will be the consequences and impacts of a society that monetizes and securitizes ever more of our thoughts and activities on human life?

Rather than worry about the future capacity of cloud storage, I think we should be worried about what impact the exponential growth in data and transactions will have on the efficiency of systems going forward.

When you buy the next application for your business and want to host that application in the cloud, where would you begin to look for cost estimates or comparisons to see what the monthly cost will be based on your business use? Often application vendors do not know what their applications will actually do for your business. Why? Because they are really good at building functionality for the business users and industry, but they are not infrastructure experts and cannot predict what your resource usage would be since each company has different workloads and growth rates. Therefore, they

24   Twitter, accessed November 28, 2022, https://twitter.com/aplusk.

25   Daniel Kolitz, "Are We Ever Going to Run Out of Digital Storage Space?," Gizmodo, May 24, 2021, https://gizmodo.com/are-we-ever-going-to-run-out-of-digital-storage-space-1846924497.

will provide you with the basic server requirements and say, "This should work for you."

The challenge most organizations encounter is that after months or a year, the server is having performance problems or is out of resources. So they add more space, thereby decreasing efficiency by using more resources to process the same requests. My goal is to change this Groundhog Day scenario by instrumenting efficient scores across the organizations for all servers, applications, and codes. If enterprises are successful in implementing controls with application development to only accept code that meets or exceeds the efficiency standard, then they will be able to reap the benefits.

So how do we bring more visibility and transparency to the millions of applications and systems out there? There is no Kelley Blue Book for applications or servers. Why? Because every application is unique, and once you apply your business model and start using the application, the resources required will vary. Even on premise, no one asks how much it costs to run an application over three to five years, as they are focused on what size of server it takes to run the application on day one.

Clients spend millions of dollars to support and enhance the applications because they were not efficient or well designed from the beginning.

Think about how long applications have been installed at your organization and what the Total Cost of Ownership has been. Was that budgeted into the original purchase price, and if so, how accurate was this estimate? What about the amount of development and administrative time your team devotes to optimizing and supporting the application because it has performance issues or other stability issues? I have seen, firsthand, clients spend millions of dollars to support and enhance the applications because they were not efficient or well designed from the beginning.

When the technology group buys the application, they evaluate it based on the functionality and decide whether it will work for their needs. Suppose you do your research and there are three applications that do the job you need. How will you decide which to purchase? You might want to know which is less expensive, which has better support, or which comes more highly recommended. You'll start to evaluate on a set of criteria—price, support, look and feel, and functionality—but we never look past the point of installation to see how efficient or what the TCO is for the application over the next five years and beyond. (See more on Total Cost of Ownership for technology in chapter 6.)

THE TWENTY-YEAR VIEW OF CODE

If your developer writes bad or inefficient code for an in-house application, that bad code is going to eventually cost the company additional resources to host, support, and troubleshoot. It could impact your customers' user experience if the code is slow, which also ultimately comes down to money.

Assuming there are no system outages or poor performance issues that impact users, this code will most likely continue to live as is in the system, using precious system resources. It could be costing the company in other ways, as there is an opportunity cost of not optimizing the code. Chances are there's a better way to standardize or influence the process to accomplish the same function and reduce the resources required, thereby making it faster and easier to support. Still, there is another dimension that I will talk about in chapter

5 that could have the most damaging cost of all—the cost to the environment.

When developers write code for applications, 99.9 percent will never know or care how many resources that code uses or how much it costs once it's in production. Most developers never know if this code is being called one time or ten million times in production, nor do they know how long the application will live—one day, ten years, or twenty years. But what if that piece of code could be 10 percent more efficient—or 50 percent? What is that code really costing us over the life of the application? What if your company writes applications that are used by other companies and your application is installed at one thousand other companies? What is the cost of that inefficient code across the one thousand implementations?

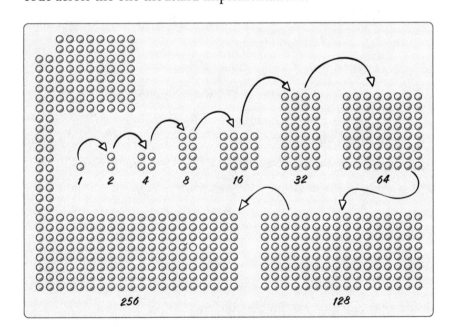

Figure 10: The complexity and compounding nature of code.

What would the impact on the customer experience be if that code could be faster? People are impatient and expect a quick response time. We are not thinking about the twenty-year view when we ask someone to build a widget for us, but we need to start measuring efficiency just in case our application is successful. We need to set higher expectations around it. In the end, both the company and users benefit from an efficient application because it is faster, uses fewer resources, and saves money.

As the amount of data grows over time, the code will use more resources to process each request, which costs more money. If we amortize and depreciate the cost of business equipment over time, why aren't we considering these metrics for applications or processes as well? Separate from data growth is the growth of your customers, a.k.a. users of the system. As the amount of data and transactions grows, the number of resources will grow at a higher rate and compound over time. Now think again—what is the cost of the bad or inefficient code over the lifetime of the application?

As everyone moves to the cloud and to a utility compute model (charging per consumption), it brings to the forefront the issue of poorly designed processes that use more resources. Why? Because the current goal for application development projects is for the code to meet functional requirements (i.e., solve a request), pass QA, and be placed into production. This model doesn't consider the bigger picture, which is where the greatest inefficiencies lie. It might cost $200 to develop the code offshore and get the job done. But if the code is inefficient, it could cost the company tens of thousands of dollars over the next twenty years. Just like MPG, each application needs a rating to determine its efficiency so we can prioritize refactoring and optimizing inefficient processes that are costing the most money every day.

DEVELOPING AN EFFICIENCY STANDARD

As the focus on technology shifts to creating greater efficiencies within systems, we need to create efficiency benchmarks to measure against a target "MPG" for applications, if you will. Since efficient applications and systems use less power, cooling, and system resources, which benefits the environment, we can make a case for a sustainability focus as well (e.g., technology groups earning green certification just as businesses do).

As discussed in this book's introduction, we've been living in a world of abundance when it comes to technology for twenty years, yet we haven't changed the way we've been looking at the health or efficiency of technology systems. There will come a time when we cannot keep adding more resources, as it will be physically challenging and cost-prohibitive to do so. Technology budgets are not infinite, and data centers, cloud or not, are not infinite.

If you are designing a brand-new application, you should be thinking about its efficiency over time, especially if you are going to be hosting and supporting it. Ask—and answer—these questions *before* you build:

- Application Questions

  - How many users will be on the application in twelve months?
  - How much data will be created in twelve months?
  - What is the estimated annual growth rate of application users and data? How will this impact the resource consumption of the application?

- Code/Process Questions

  - How many times will the main functionality (code) be called in a day?
  - How many resources (CPU, memory, storage, and bandwidth) will the code use in a day?

▫ What is the financial cost of this code for each execution? What about over the next twelve months? Next ten years?

I can envision one day when we can see the relative financial cost of code before we deploy it to production. Only then can we truly own our own destiny in a utility compute model.

Techopedia defines *code efficiency* as follows:

Code efficiency is a broad term used to depict the reliability, speed, and programming methodology used in developing codes for an application. Code efficiency is directly linked with algorithmic efficiency and the speed of runtime execution for software. It is the key element in ensuring high performance. The goal of code efficiency is to reduce resource consumption and completion time as much as possible with minimum risk to the business or operating environment.[26]

According to Techopedia, one of the recommended best practices in coding is to ensure good code efficiency. They recommend the following actions to achieve this:[27]

◆ Make use of optimal memory and nonvolatile storage.
◆ Ensure the best speed or run time for completing the algorithm.
◆ Make use of reusable components wherever possible.
◆ Make use of error and exception handling at all layers of software, such as the user interface, logic, and data flow.
◆ Create a programming code that ensures data integrity and consistency.
◆ Develop a programming code that's compliant with the design logic and flow.
◆ Make use of coding practices applicable to the related software.

---

26 "What Does Code Efficiency Mean?," Techopedia, accessed on November 23, 2022, https://www.techopedia.com/definition/27151/code-efficiency#what-does-code-efficiency-mean.

27 Ibid.

- Optimize the use of data access and data management practices.
- Use the best key words, data types and variables, and other available programming concepts to implement the related algorithm.

EXISTING MODELS OF EFFICIENCY STANDARDS

Many organizations are working to create standards to run everything from construction projects to business operations, food production, electricity, and much more. Reviewing some of the strides being made in these areas is helpful to inform the steps we need to take in the technology industry.

GREEN CERTIFICATIONS

On its website the Green Business Bureau (GreenBusinessBureau.com) lists several different types of "green business certifications," citing:

Green business certifications are … a perfect way to demonstrate a company's commitment to sustainability in a credible way. Trusted third-party organizations like LEED, B Corp and Green Business Bureau provide certification programs that can prove your company's green cred. If you're a small business owner, sustainability manager, or green team leader, you need to understand your certification options and know how they work. Not all certifications are created equal and certifications can focus on a variety of areas including products, buildings, policies, processes, and emissions. There are many programs to choose from in each area.[28]

See more about the impact of sustainable measures on the technology industry in chapter 5.

---

28 Bill Zujewski, "How to Choose the Right Green Business Certification," Green Business Bureau, April 28, 2022, https://greenbusinessbureau.com/topics/certification-benefits/how-to-choose-the-right-green-business-certification/.

ENERGY STAR CERTIFICATION

Energy Star's certification offers a path to a more efficient data center. The website SmartEnergyDecisions.com stated in 2021:

Over the past 12 years, manufacturers of computer servers have shown leadership through the ENERGY STAR® program to improve server energy efficiency and help reduce data center energy consumption. Computer servers that earn the ENERGY STAR label meet minimum power supply efficiency levels and Server Efficiency Rating Tool (SERT™) overall efficiency scores (normalized work delivered/watt) and offer enabled power management features. ENERGY STAR has informed server buying decisions for U.S. federal, state, and local government procurement and corporations by enabling the assessment and comparison of servers based on their work delivered per unit of energy consumed—a key indicator of server efficiency.[29]

This measure tests physical servers to determine efficiency levels but not the code running on the servers. The models we're suggesting would be focused on measuring the applications, or code, running on the servers.

MEASURING THE EFFICIENCY OF THE DEPARTMENT OF DEFENSE (DOD)

No entity is exempt from the quest to understand efficiency. The Defense Department is one of the world's largest technology organizations, with a budget of $740 billion in fiscal 2022 and continued

---

29   "Energy Star's Certification: The Path to a More Efficient Data Center," Energy Star, April 5, 2021, https://www.smart-energydecisions.com/energy-management/2021/04/05/energy-stars-certification-the-path-to-a-more-efficient-data-center.

investment in cybersecurity, artificial intelligence, and general IT expected to rise in future years.[30]

A June 2022 article in the government watchdog platform FCW (the Business of Federal Technology) states: [31]

The House Armed Services Subcommittee on Cyber, Innovative Technologies, and Information Systems released its mark for the upcoming defense policy bill, which calls for an independent assessment of military software and IT to determine how much money the department is losing—including in productivity—due to poorly performing software and IT systems.

A committee aide close to the process remarked:

Because the department and the military services often have what we consider underperforming, poorly performing software and IT, these service members are wasting an enormous amount of their time which is not spent … doing the things that we need them to do as a military because they're literally staring, waiting at their computer for their computer load, for their email to load, for one system to talk to another …

If we could quantify that, as many commercial companies do in terms of the cost imposed in terms of lost time, then we could have a number that we could take and illustrate that investing in things like software and IT actually will save the department money in terms of lost working hours.

---

30 Lauren C. Williams, "Why DOD Is So Bad at Buying Software," FCW, November 8, 2021, https://fcw.com/acquisition/2021/11/why-dod-is-so-bad-at-buying-software/259180/.

31 Lauren C. Williams, "Lawmakers Want to Know How Much Bad Software Costs," FCW, June 7, 2022, https://fcw.com/defense/2022/06/lawmakers-want-know-how-much-bad-software-costs-dod/367885/.

Whether an application is new or existing, essentially we need to apply an algorithm like the examples above to measure its efficiency and identify how we can make those applications more efficient.

EFFICIENCY AS A KPI

Once we determine efficiency KPIs for technology, how does this impact overall workload health? If your system is more efficient, it has more capacity, meaning it uses fewer physical resources, so the response times are faster, and it can process more transactions per server, thereby improving performance. Healthy applications are more efficient, leading to healthier workloads.

It's my belief that over the next five years, developing efficient code and applications, not just functional code, is going to be one of the biggest shifts in how we think about application development. The majority of applications remain online for ten to twenty years to support a business. Even today in my business, we frequently see applications hitting their fifteenth and twentieth anniversaries without application owners going back to refactor and optimize them. And if they're written inefficiently, that costs a lot of money in the long term and could negatively impact users.

There is currently no transparency into the efficiency of the code development process nor when it is in production. There is no forecasting out ten years in terms of capacity, performance, or costs. We are not saying every application has to be rewritten. But if we start allocating 5 to 10 percent of the development time in a year, optimizing the top 5 percent inefficient code, which, based on our averages, consumes 40 to 80 percent of all resources in the application, we can start to demonstrate some real progress toward making our systems more efficient.

My hope is that the era of abundance takes a sharp turn downward someday. Just look at the supply chain issues, staffing shortages, and worldwide inflation impacting the better of 2022. It seems like a perfect storm of inefficiency to have come out of nowhere, but if you ask some good economists, I'll bet someone saw it coming.

CHANGE YOUR ACCEPTANCE CRITERIA

We have not been looking at the efficiency of code as a major driver in our application development processes. This needs to change by including the efficiency of code as part of the acceptance criteria for the business. If we just continue to live with code meeting the functional requirements of the business, this will not be sustainable in the future. It needs to become table stakes that code is developed with a certain level of efficiency. And we need to start to measure the efficiency of systems as a whole.

Beyond changing the acceptance criteria, we need to educate developers to focus on efficiency and teach them how to write efficient code for new applications. I think this skill has been lost, as we are supporting poorly optimized, inefficient code every day at Fortified.

The design choices we make for the data in terms of coding, processes, and modeling all have lasting impacts on the bottom line, both from a resource perspective and more importantly on the financials, as most applications are in use for ten to twenty years. What is

> **The common link among everything we do, whether it's performance, financials, or the environment overall—it all comes down to efficiency.**

the Total Cost of Ownership of that code long term, and how can this be influenced during the design process? (See more about the Total Cost of Code for technology systems in chapter 6.)

Scoring efficiency starts within applications but then must track up to the overall application, the system, and someday the enterprise. Looking at the total cost of our systems from as early as when design decisions are made (the data model, code, and process design) through to the life of the application means looking not just at the financial costs to the overall system but also eventually to the greater environment. One thing I've realized in my career is the common link among everything we do, whether it's performance, financials, or the environment overall—it all comes down to efficiency.

In the next two chapters, we'll take a closer look at both the financial and environmental impacts of system inefficiency. Only by understanding the true scope of the problem can we begin to devise solutions that benefit the entire industry, if not the entire world.

KEY TAKEAWAYS

- Everyone in the industry is always focused on fires or features. No one is focused on making their application or system more efficient, but we need to because the era of abundance in technology will not last forever.
- We need to establish a measure for an efficiency KPI that can be used across the industry in much the same way that the automobile industry has a standard for fuel efficiency (i.e., MPG).
- The industry is too focused on the functionality versus efficiency of the applications, whereas the focus needs to shift to accepting code based on functionality and efficiency so the Total Cost of Ownership is lower.

- It should be a priority within the enterprise planning process to budget a minimum of 5 percent of application development time on identifying, refactoring, and optimizing inefficient processes.
- If we can develop an "efficiency KPI," this will not only help to improve the efficiency of systems now and in the long term but also reduce the resources and financial costs to our physical, as well as technological, environment.

Bringing Financial Transparency to Technology

The majority of organizations are struggling with the increasing costs of technology and the lack of financial transparency into technology services. Today if the CFO wants to know what is driving an increase in cloud costs, the only details he or she receives are down to the resource or server level. He or she cannot drill into the application or processes to determine what is costing the most. In addition to this lack of detail, there are few controls in place to prevent this situation from reoccurring since efficiency of code is a gating factor that is limiting our knowledge of true financial transparency today.

Having transparency into the true cost of your technology systems means knowing not just how much you spend from a budgetary perspective but also how much it actually costs your company to run and maintain the technology. What's more, it means understanding how much value the technology brings to your business, which in turn offsets some, if not all, of the costs to run it.

Consider the "cost versus value" woes that come with cloud-based data storage, perhaps the biggest contributor to the era of abundance over the past twenty years. A survey from 451 Research found the following:[32]

- 80 percent of finance and technology leaders acknowledged that poor cloud financial management has had a negative impact on their business.
- 85 percent reported overspending their budgets.
- 57 percent said that cloud cost management was a daily worry.
- And my personal favorite is 51 percent of respondents from finance said they overspend on cloud compared to 37 percent of the respondents from technology, which shows a real lack of financial transparency into the true costs of running systems.

To me, having financial transparency into the cost of your technology systems means knowing three things:

1. The cost of the technology
2. The value of the technology
3. The cost "takeouts" or costs of efficiency improvements that could be reduced or eliminated

Financial transparency into data costs isn't just a nice-to-have, according to Peter Wollmert, EY's global Financial Accounting Advisory Services (FAAS) leader who, in a 2018 article for *CFO* magazine, said:[33]

Organizations have an urgent need to develop reporting transparency that builds trust and helps explain how they are creating

---

32   Survey from 451 Research, "Cost Management in the Cloud Age," as reported by Apptio, *FinOps: A New Approach to Cloud Financial Management*, https://cio-institute.com/wp-content/uploads/2020/07/FinOps-A-New-Approach-to-Cloud-Financial-Management-1.pdf.

33   Peter Wollmert, "Is Everything That Counts Being Counted?," EY, November 12, 2018, https://www.ey.com/en_gl/assurance/how-digital-transformation-of-reporting-connects-trust-and-long-term-value.

long-term value by exploiting the data at their disposal and turning it into a strategic asset.

This "reporting transparency," in my opinion, cannot be accomplished without understanding the value of the technology delivering the services and data to the business.

A survey of one thousand CFOs and controllers by EY's FAAS group found that only 58 percent of respondents believed the current level of trust between the public and large companies is "high" or "very high."[34] A key culprit for the trust gap was a lack of transparency in corporate reporting—including nonfinancial reporting.

Financial information is the low-hanging fruit. We have balance sheets and other financial statements, cash flow, invoicing and purchase orders, bank ledgers, and corporate tax returns as a paper trail. We've got a host of financial software led by QuickBooks and Microsoft Dynamics built and solely dedicated to the task of financial reporting.

Wollmert goes on to say:

Organizations must account for and explain performance much more clearly, coherently, and transparently, and manage nonfinancial information with the same rigor and assurance as financial information.[35]

What's at stake without this reporting of all nonfinancial assets as well as financial? Trust, both from investors and the public, for one thing, according to the article. More importantly, so is the "business's potential for long-term value creation."[36]

So how can we get to this point of financial transparency into nonfinancial assets like code and data? In this chapter we focus on bringing financial transparency to technology systems to help us

---

34   Ibid.

35   Ibid.

36   Ibid.

understand where our spending is being optimized and where it's being wasted. This is an important and often overlooked part of the conversation that can have an enormous impact on the technology industry going forward by bringing the "rigor" and "assurance" that's required of financial results to data results as well.

THE CHALLENGE WITH FINANCIAL TRANSPARENCY

In previous chapters I discussed the importance of asking different questions—the right questions—to help determine the health of systems. The same holds true for the financial questions we ask regarding technology.

When CFOs look at the financial costs versus opportunities, they're primarily seeking to understand where to gain efficiencies, where they should be spending, and the ROI or impact of these costs on the bottom line. The answers to these questions lie within several different systems. When it comes to their technology spending, however, CFOs don't have that data centralized in one place. Often it has to be consolidated into Excel and other applications in order for the costs to be visualized and processed. So how does a CFO understand whether a technology solution or application is meeting the business milestones, as well as the financial milestones, of a project?

They can look at a project plan. They can track progress across that plan. Whether it's a data center migration or an application development project, they get meaningful status updates from project managers or the project management office (PMO). They can gauge how the project is going and whether it's on track for completion within the designated time frame and assess any roadblocks or new considerations.

But at the end of the project, when it's all complete, how do they know that the project was executed in the most fiscally responsible way?

They may have visibility into the costs of a project in terms of human resources and server spending, but how can they estimate the ROI of the project on day one of production and then know its true value again on day 365? How do they know if the code that's been developed is efficient, which will impact the Total Cost of Ownership long term? There is no easy way to get a holistic picture of the total cost of a technology project, especially once the system goes live and users are using the technology. With several projects going on at the same time within an organization, they likely will not even know where to find the details and opportunities that can uncover this information.

John Murphy, CFO of the Coca-Cola Company, believes that with the right executive sponsorship to drive the outcomes, far more useful and coordinated data services can be achieved. He brought together senior leaders from finance, marketing, commercial, and supply chain, working in conjunction with the head of platform services, to create a data governance and sponsorship structure at Coca-Cola. It consists of a formal routine in which he, along with the chief platform services officer, is a cosponsor of the initiatives that are underway to streamline, standardize, and create quality outcomes on finance data. Here is how he describes the team effort:

The power of having datasets and data systems that can integrate with each other and connect with others across the enterprise is enormous … It matters that I take the time, with the support of my team, to get into the details on specific issues and to be able to see what is needed and make sure the ultimate result is an improvement over what we had before. The same goes for just about any piece of the business where data is a key driver to better decision-making.[37]

---

37   John Lebate, "Finance at the Center of Data and Analysis at Coca-Cola," *Wall Street Journal*, March 4, 2022, https://deloitte.wsj.com/articles/finance-at-the-center-of-data-and-analysis-at-coca-cola-01646343127.

Getting team buy-in early on by enrolling enough followers who believe in the vision and feel empowered to bring it to life, having leaders who are truly focused on executing what they set out to do, and keeping people excited about the journey are key to making a system like Coca Cola's work effectively.

We don't need a large room with twenty people to talk about data; that just won't work.

—JOHN MURPHY, CFO OF THE COCA-COLA COMPANY

ELEVATING THE FINANCIALS FOR THE TECHNOLOGY INDUSTRY

When it comes to running the business, you have the top-line numbers—the costs to keep the lights on, pay the staff, etc. You may even have the cost of migrating everything to the cloud, but essentially finance hasn't kept pace with the changes in the pace of technology. The emphasis is still on function rather than form (i.e., "As long as it's doing what it's supposed to do, we're OK").

Finance hasn't kept pace with the changes in the pace of technology.

We've been in an era of abundance for a long time in technology, and during that time, most of our infrastructure was in a data center. Within the data center, most of the resources were purchased and could be used without any additional costs. The CFO and other business leaders could report on and analyze financials more easily, as it was a onetime cost. As we migrate more applications to the cloud (similar to a public utility in which you pay for what you use), we need to shift our approach on how the costs are budgeted for, managed, optimized, and reported on. Without better controls and visibility in place, CFOs and CIOs will continue to have challenging conversations about the cost and value of technology.

Once again it comes down to asking different questions—the right questions. Rather than asking, "What does it cost to build the widget?" we need to take a closer look at what it costs to develop the code and run the application in a utility model. We need to have the right financial measurements in place to determine whether it costs a dollar a day to run the code (if it was coded efficiently) or $100 a day (if it wasn't).

To help figure out the right questions to ask, business intelligence platforms have arisen to supply tools and templates to help technology professionals adjust their KPIs to reach modern-day standards for performance. Datapine offers a complete list of the "Top 20 Technology KPIs and Metrics" along with associated dashboards to help technology professionals address each one:

Total vs. Open Tickets	Do you measure the ticket churn over time?
Projects Delivered on Budget	Can you keep your budget within limits?
Average Handle Time	How do you keep your tasks under control?
New Developed Features	How many features do you continually develop?
Number of Critical Bugs	How many bugs do you regularly encounter?
Server Downtime	Do you know why and when downtime happens?
Backup Frequency	How frequently do you back up your data?
Cyber Security Rating	How is your overall security strategy?
Amount Of Intrusion Attempts	What frequency & type of threats do you face?
Mean Time To Detect	How long it takes you to detect an attack threat?
Mean Time to Repair	How efficiently you deal with unexpected events?
Phishing Test Success Rate	Are your employees aware of potential threats?
Unsolved Tickets per Employee	Do you monitor employee's effectiveness?
Reopened Tickets	Are you handling your tickets efficiently?
IT Support Employees per End Users	Do you have enough IT support?
Accuracy of Estimates	Do you estimate your team's time correctly?
IT ROI	How profitable are your investments?
IT Costs Break Down	Are you able to identify your costs' breakdown?
IT Costs vs. Revenue	Do you compare your IT expenses to your revenue?
Team Attrition Rate	Do you manage to keep talented employees?

Figure 11: Datapine's "Top 20 IT KPIs and Metrics."
Source: https://www.datapine.com/kpi-examples-and-templates/it

Only four out of twenty are centered on the financials. This is a good start but doesn't go far enough.

There are technical questions, and there are ROI questions, which are financial questions. Some of these are calculated manually, and some are instrumented. We need to move to more instrumentation of the financial KPIs and metrics and treat them on par with the technical KPIs and metrics if we want to sustain the rate of growth in data that we've been seeing over the last few decades.

CFO magazine reported:[38]

For both individual businesses and private equity sponsors, having absolute transparency into "the numbers" is a must. Some CFOs and their organizations enhance data operations for those purposes, but far too many are not. And that could be a costly mistake.

In the future a critical determinant of success will be getting smarter about converting data into knowledge. Those who continue to view data as only a cost will be passed by, according to the authors from Alvarez and Marsal.

This heightened way of thinking about data should be the most important goal of business today, with the long-term goal being to be able to assign value to data. The way to accomplish this is to start to assign value to technology, of which data is a critical component.

Starting at the bottom, we need to change the mindset when we set out to write applications or migrate processes to the cloud. If we want to obtain a financial measure from the outset, we need to look at technology services differently and ask, *What's it going to cost from the code level or data perspective to be able to support each application?* This is essential for looking toward where things will be tomorrow because the era of abundance is not going to last forever. At some point the writing of big checks for technology will come with more accountability, and those in the trenches will need to be more efficient and conscious of resource usage.

MORE ACCOUNTABILITY FOR APPLICATIONS

Here's how things work today. The building of a new application is approved within the start-up. The CIO writes a check to fund the

---

38   Steven Lee and Joey Baruch, "Data Centralization Provides Operational Insights," *CFO*, February 2022, accessed February 14, 2023, https://www.cfo.com/technology/2022/02/data-warehouse-centralization-analytics-benchmarking/.

cost of building the application, and the decision is made to build it in the cloud to save costs. The founders are happy they got their application funded, and a project charter is drawn up detailing roles and responsibilities, milestones, and expectations. Then the project team is assembled; a project plan is created.

The project plan is focused on meeting the timeline, goals, and objectives of building the application in the cloud. All of the specifications are drawn up, and the plan is handed down to the developers to design and develop the application according to specifications. The developers then start coding and testing, and when their tests meet the stated goals, they promote the code and move on to the next task.

The project manager keeps track of resource costs in a project plan. Costs associated with the development environment are tracked through some type of project management software (such as Jira, Wrike, or Asana) while another cost management software tool is tracking project costs. Somehow (probably manually) the data and the cost information are merged into an Excel document for interpretation. This "data" is then fed up to the CIO.

If at any point in time someone within the tech stack wants to know the status of the project, they could check the software tool and determine if there are any roadblocks. They can look at the hours put in on the project because these are logged into yet another system.

OK, so we can understand how many hours went into building each feature. But how is success defined? How can we know how much it will cost to run this application in production? How do we know how efficient the process is and what the total cost of the code will be? How do we truly know the costs that are realized apart from the project costs (i.e., the cost to maintain and support the system rather than the cost to build it)? Where are these budgeted for and tracked? For example, can we put a cost against workload friction? Against workload health? We don't have any of these operational data points that are dependent on the infrastructure and timing of other processes and requests executed in the same environment once the codes are in production.

Every organization has servers. Some have excess capacity, some not enough. Some have resource constraints (friction, traffic jams, etc.). We do not see any financial data on these factors today, which means we do not have a Total Cost of Ownership for the application. The more important question is, *What is the Total Cost of Ownership over five years, and how is this data being communicated to the CFO?* At the same time, everybody's looking at too much data that's not telling us the things we really need to know.

Before the decision is made to push something to production, what if we could know the Total Cost of Code when it's launched? Once it is launched, how do we forecast this out annually for budgeting purposes? Then how do we identify the potential cost takeouts because if the code can be 30 percent more efficient, this would further reduce the cost? The CFO is not the only one who needs this data; everyone from the developers to the management does.

> **Everybody's looking at too much data that's not telling us the things we really need to know.**

At Fortified we're writing algorithms that allow our team and clients to see the Total Cost of Code along with other metrics. Just like a car, every server has an oil level (capacity) and a battery (to power the engine) and is run at a level of speed that either will tax the system or is sustainable. We check these regularly as a matter of course for our cars; why not for our technology?

We need to be able to instrument financial data metrics in the same way so we can determine exactly how much the piece of code we wrote to perform the task is going to cost—in terms of resources, staff, capacity, and several other indicators that all feed into the Total Cost of Ownership for the code or application. We need to develop a system or set of processes that allows us to have a financial view of our technology investments. We need to make it easy to access the cost data from both the operational and financial systems that analyze and report data in a way that people can understand not just what the costs are today but also where the cost takeouts are and the value of that technology to the business. This way when prioritizing fixes or application features, the process won't be random, leaving us to hope for the best in terms of cost savings, but instead financially driven. It will be calculated so we can make assessments of the value gained from making the fix versus the costs incurred for not making it.

GARTNER: RUN TECHNOLOGY LIKE A BUSINESS

Technology research and consulting firm Gartner offers this advice to CIOs: run technology like a business and develop a framework for making long-term changes on the financial side. To accomplish this requires shifting budget planning from the traditional asset-based view of hardware, software, head count, and outsourced services to also include:[39]

---

39   Laurence Goasduff, "6 Pillars of IT Financial Transparency," Gartner, July 19, 2017, https://www.gartner.com/smarterwithgartner/6-pillars-of-it-financial-transparency.

38   Luke Dormehl, "When We Run Out of Room for Data, Scientists Want to Store It in DNA," Digital Trends, July 8, 2018, https://www.digitaltrends.com/cool-tech/dna-data-catalog-startup/.

39   IDC's Data Age 2025 study, sponsored by Seagate, April 2017, as reported by Digital Trends, July 8, 2018, https://www.digitaltrends.com/cool-tech/dna-data-catalog-startup/.

40   "Energy Demand, Three Drivers," ExxonMobil, October 5, 2022, https://corporate.exxonmobil.com/Energy-and-innovation/Outlook-for-Energy/Energy-demand.

- *the technical view of servers, storage, and networks*, which enables analysis and benchmarking of technology spend on a per-unit cost basis;
- *the business services view*, which communicates the cost of the services that the technology department provides in terms that nontech people can understand; and
- *the investment view*, which distinguishes the amount of money spent on investments in new capabilities and their potential value from the costs of doing everyday business.

Gartner also offers these steps a CIO can take to promote greater financial transparency:

- Track both the actual and the forecast project cost through to completion. Ensure that business leaders understand the charge-backs to their units to minimize complaints and dissatisfaction.
- Incorporate benchmarking into the annual planning process to provide financial transparency, as well as a mechanism to identify areas of opportunity.
- Establish a baseline for technology spending and implement a strategy to optimize, execute, and track ongoing progress (rather than react to immediate budget cuts and then return to business as usual).
- Rather than ensure the "trains are running on time," demonstrate the business value of technology through transparent communication via a constant feedback loop.

You may be reading this and shaking your head. *No, that's not the case at our company.* Good! I want to hear what you are doing that works. I want to share best practices among the community. I want things to change for the sake of the future of our industry. (Join me at *bendebow.com/contact!*)

THE FUTURE OF WORKLOAD ANALYTICS: A VISION

A retirement calculator will ask for your age, income, savings, and preferred lifestyle to make a determination of how much money you will need to live your dream life at retirement, along with recommendations on how to amass those funds. Only you can decide how willing you are to forgo certain indulgences now to achieve the life you desire in the future.

Google Maps lets you put in your destination and then analyzes a multitude of traffic patterns in an instant to fire up three routes you can take—one can save you time, one can save you turns, and one is more scenic. Only you as the driver can decide which route is optimal for you based on your requirements at that given time.

Can you see where I'm going here? What if we could have a tool or application that works like a retirement calculator or Google Maps? Put in your code and your business goals, and—magic—you see the costs today and for the future along with alternative options based on different assumptions for optimizations, cost takeouts, and growth metrics. It's up to you to choose the optimal plan and then constantly check the application for any deviations and alternative routes to the desired outcome as your business grows. Then based on your adjustments, the system can learn what you need and start to serve these options up automatically.

If instead of three routes to drive or a dollar amount to save for retirement being served up, we get these options:

- process things at different times using dynamic scheduling (to optimize capacity)
- optimize the size of the server and remove the contention in the system (to remove friction and rightsize the system)
- remove these worst-performing (least optimized) pieces of code (that are eating up 50 percent of your overall budget)

And so on. This is the vision for instrumentation to bring financial transparency to technology systems.

It won't matter how much financial transparency we have into systems if those systems are no longer sustainable. In the next chapter, we investigate the impact technology is having on the natural environment and what steps are being taken to reduce its carbon footprint. If you're thinking, *This is a chapter I can skip*, I urge you to reconsider. Some may argue it's too late for climate change, but I believe technology has an important role to play in protecting our future and that there's still enough time for us to make a real difference.

KEY TAKEAWAYS

- Having financial transparency into the cost of your technology systems means knowing the cost of the technology, the value of the technology, and the cost takeouts, or savings gained by efficiency improvements. (See more on this in chapter 6.)
- We need more instrumentation of the financial metrics and to treat them on par with the technical metrics if we want to sustain the rate of growth in data that we've been seeing over the last few decades and achieve a more accurate budget.
- Until we can understand the cost to maintain and support systems in addition to the cost to build them, we will not have any data on their Total Cost of Ownership.
- We need to budget and forecast out the Total Cost of Ownership for one, three, and five years across all applications and compare this against the actual costs.
- As there will be gaps between the budgeted and actual costs, finance, infrastructure, and the developers need to work together to align the costs.

CHAPTER 5

Can We Afford to Ignore the Environmental Impact of Technology?

I f you think this is a chapter about the buzzword of the past two decades—sustainability—and therefore skippable, I urge you to reconsider. I am not a "green" guy but can be found in nature almost every day. My teenage daughter is educating me on how we're polluting our oceans and we need to start doing things differently if we want to save the environment. She's learning about it in grade school. And she has a good point. We all need to think, *How can I help and start making a difference?* Every individual contribution matters, even if it's a small one.

The same holds true for optimizing technology's role in conserving resources and protecting the environment. Much has been written about the role of technology to create sustainable solutions for just about every industry. It's top of mind as new ESG (environmental, social, and governance) standards permeate boardrooms as a must-have, and investors consider ESG ratings as part of their evaluation criteria.

Every company needs a sustainability pledge, at a minimum, and a list of KPIs for tracking their progress. This is good news; we need to save the planet, and technology can help. But what's missing from this conversation is *the degree to which technology is contributing to the environmental crisis the world is facing.*

One of Gartner's "Top 10 Strategic Tech Trends for 2023" is sustainable technology. By 2025 they predict that 50 percent of CIOs will have performance metrics tied to the sustainability of the technology organization.[40] Here's why:

Sustainable technology is an area that has risen to the top of priority lists for many company executives and should be looked at as a framework of solutions that increase the energy and material efficiency of IT services, enable sustainability of both the enterprise and its customers, and drive environmental, social, and governance (ESG) outcomes. Through the use of technologies such as artificial intelligence, automation, advanced analytics, and shared cloud services, among others, companies can improve traceability, reduce environmental impact, and provide consumers and suppliers with the tools to track sustainability goals.

ENVIRONMENTAL IMPACT OF TECHNOLOGY

Consider these findings:

* Each day around 2.5 quintillion bytes of data are created, courtesy of the 3.7 billion humans who now use the internet.[41]
* Ninety percent of the world's data has been created in the last two years (as of May 2013).[42]

---

40   Peter High, "Gartner's Top 10 Strategic Tech Trends for 2023," *Forbes*, October 19, 2022, https://www.forbes.com/sites/peterhigh/2022/10/19/gartners-top-10-strategic-tech-trends-for-2023/?sh=51322d1a4cb4.

41   Dormehl, "When We Run Out of Room for Data."

42   "Big Data, for Better or Worse: 90% of World's Data Generated over Last Two Years," Science Daily, May 22, 2013, https://www.sciencedaily.com/releases/2013/05/130522085217.htm.

- By 2025 it's projected that the world will create two hundred zettabytes of data—and it's estimated that 50 percent of this will need to be stored in the cloud.[43] (You do the math.)

- Global energy demand is projected to reach about 660 quadrillion Btu in 2050, up ~15 percent versus 2019, reflecting a growing population and rising prosperity, according to the December 2021 *Outlook for Energy* report from ExxonMobil.[44]

- It's estimated that in the United States alone, data centers consume about seventy-three billion kilowatt-hours of electricity each year, about the same amount of energy that six million homes consume in one year.[45]

- E-waste (the disposal of electronic and electrical equipment) has grown from 5.3 kg per capita to 7.3 kg per capita between 2010 and 2019, but recycling hasn't kept up. Further, the 2020 United Nations *Sustainable Development Goals Report* states:

The [e-]waste is mostly handled by the informal sector through open burning or acid baths, both of which pollute the environment and result in the loss of valuable and scarce resources. Moreover, workers and their children, who live, work, and play on those sites, often suffer severe health effects.[46]

I don't reference these statistics so you could consult your online dictionary and look up words like zettabyte, quintillion, or quadril-

---

43   Steve Morgan, "The World Will Store 200 Zettabytes Of Data by 2025," *Cybercrime*, June 8, 2020, https://cybersecurityventures.com/the-world-will-store-200-zettabytes-of-data-by-2025.

44   "Energy Demand: Three Drivers," ExxonMobil, August 28, 2019, https://www.exxonmobil.co.uk/Energy-and-environment/Looking-forward/Outlook-for-Energy/Energy-demand.

45   Arman Shehabi et al., *United States Data Center Energy Usage Report* (Berkeley, CA: Lawrence Berkeley National Laboratory), 2016, LBNL1005775.

46   United Nations, *The Sustainable Development Goals Report 2020*, p. 12, https://unstats.un.org/sdgs/report/2020/goal-12/.

lion. I think you get the idea—the carbon footprint of technology is big, really big. It's like my view on winning the lottery—once you get past $25 million or so, it doesn't really make a difference how large the pool is.

Seriously I reference these statistics to impress upon the community that this idea of endless technology resources is frivolous and entitled. We are in an era of abundance, and as a result, we've gotten a bit spoiled. Even if the resources (i.e., compute, memory, or storage capacity) don't actually "run out" in our lifetime, it doesn't mean we should treat them as infinite. It just seems wrong from an ethical perspective, that is, if all the research and documentation make your head spin like it does mine.

MORE THAN MOORE'S LAW

Moore's law is a generally accepted principle in technology brought forth by Gordon Moore, the former CEO of Intel, in 1965, which postulates that the number of transistors on a microchip doubles every year, while the cost of microchips is essentially halved, creating great economic efficiencies. However, it's also widely recognized that Moore may not have anticipated just how much data growth there would be to fill those microchips and that computers would reach the physical limits of Moore's law at some point in the 2020s, a.k.a. any day now.

While Moore's law is slowing down as transistors become smaller and data becomes bigger, a new set of problems, namely, the heat they generate and the power they consume, is now consuming the industry. According to Investopedia:

The high temperatures of transistors eventually would make it impossible to create smaller circuits. This is because cooling down the transistors takes more energy than the amount of energy that already passes through the transistors.

As a result, several start-ups are working to create a chip that defies Moore's law or at least give us a way to keep storing more without relying on Moore. Innovation is good. So we can wait around for someone to invent the next great micro-microchip that will allow us to store even more information, or we can think about how to reduce our reliance on the era of abundance supplying us with exponential data growth for an infinite amount of time.

GREEN DATA CENTERS AREN'T GREEN ENOUGH

The truth is cloud computing is just another data center owned and managed by somebody else.

—BEN DEBOW

Data centers are built to store data, and as data becomes more prolific, the way we house it continues to evolve. At its core a data center contains servers, storage, and networking, all of which need to be licensed just to support the software. All this investment in technology allows the company's applications to run so that data can be analyzed, visualized, stored, moved, or modified. Data centers require an enormous amount of electricity to power their operations.

As data grows exponentially, companies like Microsoft, AWS, and Google are designing green data centers using 100 percent renewable energy so that the data are more off-grid—the electricity grid, that is. These data centers are being created according to standards for design, building, and operations similar to LEED (Leadership in Energy and Environmental Design) certifications for green buildings.

Leading projections put global cloud revenue at anywhere between $277 bn and $328 bn for 2021 as Jake Ring, founder of GIGA data centers, wrote on DataCenterDynamics.com in 2019.[47] At that time he said:

The cloud, whether public or private, is housed in data centers and these facilities must continue to evolve if they are going to support the aforementioned predictions. It's also important to note that evolution, in and of itself, does not ensure survival for all technologies. For data centers to properly evolve so that all types of organizations can benefit, they must also become more cost-effective and efficient to operate, be more flexible to support higher power configurations (such as hyper-converged), and scale faster.

> **We should have a rating system for green-certified applications that enables companies to see and measure their green footprint for existing applications.**

I would take this thinking a step further and say we should have a rating system for green-certified applications that enables companies to see and measure their green footprint for existing applications and helps them identify what they can optimize within the application to reduce their carbon footprint.

Think of the options we have for green cars. The manufacturer takes a stand on fuel efficiency and design protocols that comply with that stand, even if it costs more because they know consumers are willing to pay more if they can save money down the road with a more

---

47   Jake Ring, "Death of the Data Center? Not According to the Analysts' Cloud Forecasts," Data Center Dynamics, February 4, 2019, https://www.datacenterdynamics.com/en/opinions/death-of-the-data-center-not-according-to-the-analysts-cloud-forecasts/.

efficient car. Plus, they're doing their part for the environment. How can we get technology providers to take a stand on the environment and make them stick to it?

Or take the hotel industry as another example. When you check into a hotel, you can choose the "green option" (i.e., they won't clean your room or give you fresh towels every day to save water and electricity). What if we had the option within green data centers to choose the level of "green" we want for our applications to ensure that they're rightsized? How can we make the applications that go into the data centers more efficient to create more bang for the buck with green data centers? For example, the goal of most nonproduction servers is testing functionality, not performance. If this is true, we should be running nonproduction on slower, smaller servers that use fewer resources. This would save a significant amount of resources and costs, considering most environments are two-thirds nonproduction.

Green data centers and servers are fine, but if the application uses twenty million CPU cycles to process a simple question against a large amount of data and that application is called ten thousand times a day, how can we claim these operations are truly green? It promotes a false sense of comfort to say the data centers are green and therefore saving energy if the applications themselves are inefficient and wasteful. To be truly green, we need to certify the applications as well as the data centers that store them.

USING TECHNOLOGY INNOVATION FOR GOOD

The good news is many sources agree that technology can play a role in lessening the pending environmental crisis. The bad news is, to quote the World Economic Forum, "technological innovation … it's complicated."[48]

"We've seen substantial growth in investment in technology R&D—everything from expanding mobile access to artificial intelligence (AI)—up to $2.2 trillion in 2017 from $1.4 trillion in 2010 and $741 billion in 2000." And there are innovations every day to show for it, particularly with COVID-19 accelerating R&D in healthcare technologies.

A 2020 article from the *Production and Operations Management* journal from Wiley, titled "Information Technology and Sustainability: Evidence from an Emerging Economy," found that[49]

green IT investment is positively associated with a higher profit impact and that this association is partially mediated by a reduction in IT equipment energy consumption. In addition, we find that operations-oriented green IT implementation is positively associated with both IT equipment energy consumption reduction and profit impact, whereas supplier-oriented green IT implementation is positively associated only with reduction in IT equipment energy consumption.[50]

The tech industry titans are certainly doing their part to reduce energy consumption both for themselves and their customers.

MICROSOFT

Microsoft has taken considerable measures to reduce its carbon footprint by ensuring their cloud is as much as 93 percent more

---

48 Carla Tardi, Marguerita Chang, and Amanda Jackson, "What Is Moore's Law and Is It Still True?," Investopedia, July 17, 2022.

49 Jiban Khuntia et al., "Information Technology and Sustainability: Evidence from an Emerging Economy," Wiley, November 17, 2017, https://onlinelibrary.wiley.com/doi/10.1111/poms.12822.

50 "Tech for Good: What Are the Challenges of Making Technology and Digitization More Sustainable?," World Economic Forum, September 20, 2022, https://www.weforum.org/agenda/2020/09/what-are-the-challenges-in-making-new-technology-more-sustainable/.

energy efficient and 98 percent more carbon efficient than on-premise solutions. To achieve this they plan to concentrate on four key areas: [51]

- Operational efficiencies
- Equipment efficiencies
- Data center infrastructure efficiencies
- Renewable energy purchases

In June of 2022, Microsoft introduced a new sustainability solution called Microsoft Sustainability Manager, which helps companies record, report, and reduce their environmental impact by [52]

- improving data collection to measure, report, and monitor emissions;
- transitioning to a more efficient technology option;
- migrating to the cloud to drive energy efficiencies;
- adopting sustainable technology;
- optimizing operations to decrease environmental footprint; and
- driving efficiencies and transitions to clean energy.

GOOGLE

Google is the largest corporate buyer of renewable energy in the world. Their energy purchases as of 2019 are projected to increase Google's existing renewable energy portfolio by more than 40 percent. Google Cloud is at the forefront of rapidly transitioning the entire

---

51   Microsoft, *The Carbon Benefits of Cloud Computing*, 2018, https://info.microsoft.com/ww-landing-Carbon-Benefits-of-Cloud-Computing.html?lcid=EN-CA.

52   Microsoft Sustainability, accessed November 25, 2022, https://www.microsoft.com/en-us/sustainability/cloud?.

global economy to clean energy with a commitment to have their data centers supplied 24/7 with carbon-free energy by 2030.[53]

Some of Google's notable recent achievements toward this end are the following:

- A system that can shift flexible computing tasks to times when power on the grid is cleanest
- AI technology that helps reduce the energy they use to cool their data centers by 30 percent
- Deploying smart temperature, lighting, and cooling controls to further reduce the energy used at their data centers
- Using machine learning to optimize how wind farms deliver power

Further, in its article "Best Practices for Compute Engine Regions Selection," Google suggests considering a "carbon-free energy (CFE) percentage" as a factor in choosing where to deploy compute engine resources, citing:[54]

It's common for people to deploy in a region where they are located, but they fail to consider if this is the best user experience … Multiple factors affect where you decide to deploy your app.

Until their 2030 goal is achieved, Google Cloud regions are supplied by a mix of carbon-based and carbon-free energy sources every hour and measured by a CFE percentage. For new applications running on Google Cloud, they suggest developers incorporate carbon impact into their architecture decisions.

---

53   Urs Hölzle, "Announcing 'Round-the-Clock Clean Energy for Cloud," Google, September 14, 2020, https://cloud.google.com/blog/topics/inside-google-cloud/announcing-round-the-clock-clean-energy-for-cloud.

54   "Best Practices for Compute Engine Regions Selection," Google Cloud, accessed November 25, 2022, https://cloud.google.com/solutions/best-practices-compute-engine-region-selection.

CFE percentages are published for Google Cloud regions, with Finland being the best at 94 percent and Singapore being the worst at 4 percent. Choosing a region with a higher CFE percentage means that, on average, your application will be powered with carbon-free energy at a higher percentage of the hours that it runs, reducing the gross carbon emissions of that application.[55]

As we work towards our 2030 goal, we want to empower our customers to leverage our 24/7 carbon-free energy efforts and consider the carbon impact of where they locate their applications.

To power each Google Cloud region, Google uses electricity from the grid where the region is located. This electricity generates different degrees of carbon emissions, depending on the type of power plants generating electricity for that grid and when Google consumes it.

These are certainly steps in the right direction, but the reality is Microsoft and Google and all the other tech giants can design and improve their data centers all they want; however, most applications were designed within the era of abundance and therefore will continue to consume an exorbitant number of resources, even in a green data center. So they're not solving the problem completely.

WHAT IS BEING DONE TO REDUCE THE ENVIRONMENTAL IMPACT OF TECHNOLOGY?

From innovative start-ups to increasing regulation to new standards, there is hope on the sustainable technology horizon.

---

55 "Carbon-Free Energy for Google Cloud Regions," Google Cloud, accessed November 25, 2022, https://cloud.google.com/sustainability/region-carbon.

CATALOG: MAKING DNA IN A LAB

A start-up called Catalog is working to solve the data storage conundrum by encoding data into a synthetic form of DNA. Besting conventional storage media like flash drives and hard drives, which do not have the longevity, data density, or cost efficiency to meet the global demand, Catalog is building the world's first DNA-based platform for massive digital data storage and computation.[56]

According to Catalog's website:

DNA can store millions of times more data in the same volume as conventional solutions, can last for thousands of years, and gives you the ability to physically own your data—even massive amounts of it.

It may take a zettabyte's worth of time to actually move the world's collective data storage to DNA, but kudos to Catalog and their investors for trying to find a solution and save precious resources. And I thought the flash drive was a big deal.

ESG IS THE NEW BLACK

Accenture Research has found that corporate leaders increasingly understand the need to effectively measure the impact of environmental, social, and corporate governance on their business, but many struggle to take the appropriate action. Of note are as follows:[57]

- Forty-seven percent of corporate leaders have created key performance indicators and have sourced the relevant data to allow sustainability value creation to be tracked, traced, and managed.
- Only 26 percent of finance leaders agree they have clear, reliable data to underpin each ESG key performance indicators.

---

56   Catalog, accessed November 25, 2022, https://www.catalogdna.com.

57   Peter Lacy et al., "Measuring Sustainability. Creating Value," Accenture, https://www.accenture.com/us-en/insights/strategy/measuring-sustainability-creating-value.

Many of these environmental KPIs are measuring what can be tracked today, which is a good starting point. But these measurements are overlooking the tech debt and inefficiencies of thousands of applications. It's the classic example of *they don't know what they don't know.* Therefore, they are not factoring in how much greener they can be, or money they can save, by becoming more efficient.

So how can leaders up their ESG game? According to Accenture, having access to the right data to make better decisions at every level is critical. Cloud and platform providers play a pivotal role here in tracking and providing interoperability throughout the value chain.

Accenture offers a ten-question diagnostic to assess a company's progress on sustainability efforts and measure their company's "sustainability DNA." The seventh question in the diagnostic is "To what extent do you agree that you/your organization harness(es) the potential of technology to solve critical challenges without creating harmful side-effects?" What a loaded question! I don't see how it can be answered on a sliding scale. I think we need another diagnostic to assess whether the technology is creating harmful side effects and how to quantify them.

The types of questions that would be on this diagnostic are as follows:

- How much data are you generating per year?
- What is your average data growth year over year?
- What are your technology resource costs (compute, memory, and bandwidth) per year?
- What is the growth rate per year for each technology resource needed to support the business, application, and data growth?
- How much are inefficient code or inefficient applications costing you?

- How much could you potentially save by creating more efficient code?
- How much savings will this translate to over one month? One year? Five years?
- Which of the following strategies are you using to reduce your technology's carbon footprint?

  ▫ Shifting to a data center or the cloud with green initiatives
  ▫ Evaluating capacity to determine where there is excess and where there are deficiencies
  ▫ Implementing operational efficiencies, such as shifting workloads and balancing resource consumption
  ▫ Implementing equipment efficiencies like buying infrastructure with a lower power consumption
  ▫ Making renewable energy purchases or increasing the end of server life from three to five years
  ▫ Putting restrictions on the purchases of new hardware
  ▫ Designing and building applications and processes that are efficient and use fewer resources

As you may have guessed, this last one is my area of specialty. Think of carbon-efficient applications this way: it's not just one app running on one server. If every app running on every server adopted a sustainable mindset, think of the overall efficiency savings and the "capacity" that technology would have to focus more on the good it can do rather than the limited resources it's eating up. What if each unit of savings was a credit or something measurable or standardized (i.e., a benchmark)?

REDUCING THE CARBON FOOTPRINT OF APPLICATIONS

Asim Hussain is a green developer relations lead at Microsoft. Here are his recommendations on what to measure so you can reduce the carbon footprint of your applications, which, as it turns out, also align with the same metrics to make code more efficient: [58]

- **Time to interactive.** The lower the better.
- **Page weight.** The lower the better.
- **Average server response time.** The lower the better.
- **Cost of your services.** The lower the better.
- **Utilization of your servers.** The higher the better.

We need to focus on actions in every area of our technology environment from efficiency metrics (availability, capacity, performance, stability) to financial transparency indicators. This holds true for sustainability efforts as well. For the environmental impacts to work, we need to show that if we tune something, it saves us x amount of carbon credits.

In his post on Medium, technologist Sanjiv Khosla agrees: [59]

There are no standard industry benchmarks that can rank peers on energy efficiency. For example, we should compare the energy consumption footprint of Amazon, Shopify, and Etsy across several metrics like the number of transactions, items sold, shipments processed, goods delivered, GMV, revenue generated, etc. Appropriate

58 Asim Hussain, "How to Measure and Reduce the Carbon Footprint of Your Application," Microsoft Industry Blogs – United Kingdom, accessed November, 24, 2022, https://cloudblogs.microsoft.com/industry-blog/en-gb/technetuk/2021/10/12/how-to-measure-and-reduce-the-carbon-footprint-of-your-application/.

59 Sanjiv Khosla, "Green Data Centers Aren't Enough to Exhibit Environmental Consciousness," Medium, November 7, 2021, https://medium.com/applied-software-factory/green-data-centers-arent-enough-to-exhibit-environmental-consciousness-3dec40c996af.

environmental organizations must reward companies that hit their KPIs with the smallest energy footprint.

If organizations continue to focus on rightsizing their servers, this will only achieve small gains in reducing the overall resource footprint, although they may derive financial value in reducing their licensing costs. However, the biggest gains can be achieved by optimizing their applications and processes. In our experience at Fortified, if you were to optimize the five highest consumers of resources on a typical database platform, this would yield an approximate 40 to 80 percent gain in efficiency.

If you were to optimize the five highest consumers of resources on a typical database platform, this would yield an approximate 40 to 80 percent gain in efficiency.

Just by shifting our mindset to focus on this type of approach, data and transactions can continue to grow exponentially, and we wouldn't need to continue to build new data centers or increase compute at the same pace because we'd be realizing a larger savings.

Compute is one of the greatest consumers of resources in the whole technology environment (not to mention the exorbitant costs paid to vendors for licensing applications by CPU and how this expense grows as CPU grows). CPUs run hot; therefore, they need a lot of cooling and consume a lot of energy. We need to consider the cost in terms of the financials, as well as the resource footprint as it relates to these processors.

We're in the abundance era; we expect something will be there because it's always been there. But look at the water crisis in South

Africa or the gas crisis in our own country in 2022. Resources can run out or become too expensive to rely on.

————————

Michael Ziegelheim has over twenty-five years of success in leading information technology operations for several companies including Dun & Bradstreet, Philips Electronics, WebMD, and Dynarex Corporation, a $100 million medical device manufacturer. I asked him to what extent sustainability issues impacted his daily work. He supported my era of abundance theory by saying:

We've gotten used to upgrading everything every three years: upgrade this service, this hard drive, this router, this hardware, this Wi-Fi, this modem. And everyone needs the latest laptop or smartphone to do their jobs better. While some companies may think about what happens to all the old computers, their efforts at recycling or donation have been a drop in the bucket when compared to the huge rate of equipment throughput. It's phenomenal how much wasted equipment there is ending up in some landfill.

Leaders see it as a cost of doing business to give their employees a new laptop every three years. And it's tough to fight back because in tight job markets, talent can be persuaded with shiny objects from other employers. There are certainly people and organizations who need the latest and greatest tech available to perform their jobs. However, the individual computing needs of a large majority of users can be met with technology that has been available for years. This is even more true as processing has increasingly moved to the internet and cloud.

It's cloud service providers like Amazon, Microsoft, Google, and others that are investing in the latest tech, as it's on their platforms that the bulk of computer processing is performed. The desktop computer becomes more of an appliance. This model is not going to change any time soon. I'm sure there will be new things coming that no one's thought of yet. Some may well become ubiquitous. The cloud model is here to stay. It's just what you do with it that might be new.

But when you talk about sustainability, I think of how much waste there is in the system and, more importantly, the fact that businesses aren't doing enough about it.

But even for personal use, our computing has become cloud-centric. And while I can't deny the amazing advances in mobile technology like image and video processing, which is just stunning, you have to wonder how much landfill space is continually taken up by discarded mobile devices.

ACTIONS TO TAKE NOW

Now that you've made it to the end of this chapter, I'd like to first say thank you for reading. I'm going to recap with some simple steps to take now, if any of this information resonated with you:

- *Evaluate and adjust your scheduling* to optimize the processes to run when you have excess resources, which would reduce your peak resource utilization.
- *Design and optimize code and processes* to be more efficient, which will reduce the physical resources required. This one change could provide the world with significant savings if every organization could optimize the top 5 percent of their processes over the next several years.
- *Consider the consequences of inefficient code over time.* If the data set is small now, it's not as much of a concern, but five years and ten billion records from now, the impact of that schema or application design decision on the environment could be devastating.

- *Before choosing a data center or cloud provider, review the sustainability and green scores* and then make a decision based on the impact to the planet as well as your pocketbook.

- *Implement or review ESG practices, establish KPIs to track the progress of environmental efforts,* and create your own diagnostics to continually assess your progress.

- *Implement data life cycle management,* a process for keeping track of data throughout all stages—from collection/creation to archive/destruction—as this is an easy way to manage the amount of data your company stores while balancing the number of resources required to ask questions against the data.

- *Reconsider the impulse to replace servers every three years;* do your due diligence with providers and make a judgment call to extend their life to five years based on your anticipated needs. Try running noncritical environments on older hardware that can be offline for days while you buy a replacement server.

Achieving true technology health means improving the efficiency, financial transparency, and environmental impact of technology systems for a sustainable future. Now that we've established the need for each of these measures, the next few chapters will take a deeper dive into actions we can take to bring us closer to the vision I've laid out in this book.

KEY TAKEAWAYS

- Technology has a serious and substantive role to play in addressing worldwide sustainability efforts.
- Shifting to green data centers is a start, but it's not enough. We also need to measure how to make each of our applications more efficient and therefore greener.
- Technology behemoths like Microsoft and Google are taking steps to do their part for the environment and help their customers do theirs as well. At the same time, start-ups, consultancies, and ESG rating agencies are sounding the alarm for technological sustainability.
- The importance of evaluating the carbon footprint of our apps cannot be overstated if we truly want to play a role in protecting the environment.
- It's time for every company to review their own operations and see what they can do to create efficiencies and save resources to help the planet. Smart small and think about how we can make a difference.

What Is Your Technology Really Costing You?

I was talking with a new client of ours who was telling me about an unfortunate experience he had with another vendor. He said the vendor told him that if he built their system with the infrastructure resources they advised, it would last them three years. It barely lasted one year before they had to add more resources. Specifically my client wanted to know how Fortified could help him get a better understanding of what this application needed to run for the next three years. How could he plan for future capacity while optimizing the performance?

These were his exact words:

I met with that vendor today, and they still don't know where these memory leaks are coming from and what's causing the application to go down. We have to reboot this thing proactively every week, and soon this will be twice a week because it's so unhealthy. I don't have the visibility into why this application's behaving like it is, and it's unfortunate, but ultimately the responsibility's on me; I'm responsible for making sure this application is there to run the business.

In my business we're focused on measuring the impact of rightsizing capacity (discussed in chapter 2). We're also focused on optimizing each application or process that is deployed to production, not just when it is deployed but also over the life of the application.

I told my client what he should have asked that vendor who sold him the application and what you should be asking every time when you release a new application or process: *What is the Total Cost of Ownership of the application, code, or process?*

OPTIMIZING THE COST OF APPLICATIONS

Applications are designed to make business processes simple for the technology user—at least that's the goal. Behind these applications there is much going on. It takes a lot of resources and complexity for an application or system to provide an answer, even if the response time is only a few seconds. Now multiply this by thousands or millions of system requests per second across thousands of servers across your enterprise. It's easy for things to become out of reach with so much going on simultaneously, and this relates to costs as well. If an application is supposed to last three years but only lasts one, what is the true cost of that application?

As mentioned throughout this book, everyone is so focused on the here and now, ensuring the system gets through each day without an outage. No one is thinking, *Could we do what we are doing faster, better, smarter (i.e., more efficiently)?*

In addition to optimizing applications or processes, there are two areas that impact the true costs of systems over time, or *Total Cost of Ownership*, when it comes to technology systems—the code and data.

TOTAL COST OF CODE

What is your Total Cost of Code?

As discussed throughout this book, a healthy system requires fewer resources to manage the same workload than an unhealthy system. Optimizing code and processes frees up even more resources not just today but also long term, as code does not run just for one day. In Fortified's experience virtually any system has the potential to realize capacity rationalization by at least 30 to 40 percent, and code optimization may provide another 40 to 80 percent of cost savings. This means the same workloads can be run on smaller machines, reducing cloud and licensing costs. The value of these savings is not simply short term, but over the years, the application is in use. It's not just a matter of the bottom line but it's also the consideration of what could be done with this freed-up capital to further business KPIs.

Code optimization can have dramatic impacts, and it's been my experience that the market is generally unaware that such inefficiencies exist within their platforms or within their application code unless they are forced to cut x amount from their budget or scale quickly due to a sudden influx of customers. That's when they must open the hood and see what's there.

TOTAL COST OF DATA

What is your Total Cost of Data?

Each day the rate of data creation and data acquisition is increasing across the world. In most organizations the growth rates are double digit, which is impacting the bottom line. In addition to the standard costs incurred to store your data, there are many other soft costs required to manage, maintain, query, and optimize the data. The impact of data growth on data storage is an often-overlooked component of applications that also drives some of the costs along with performance impacts.

When most people think of data, they're essentially thinking about a filing system. Each new piece of information gets cataloged and stored in a digital file cabinet. But there is a cost for storing data, and as the number of business transactions grows, data grows, and it becomes more costly to manage and maintain the data.

In technology we have storage tiers—hot, warm, and cold. Hot storage has the fastest disks and carries the highest costs. You could store data more cheaply, but it would take longer to get the information you need when the application requests it, just as if you had a closet filled to the brim with clothing and were looking for one particular sweater. The more data you have, the more this impacts performance. There is a trade-off between speed (efficiency) and cost.

ARE YOU A DATA HOARDER?

One Fortified client designed their system back in the late nineties, and they haven't archived any data. Their Oracle database has thirty *terabytes* of data that cost money and consume resources to manage, maintain, optimize, and back up every night. Every byte within every terabyte every night, and most organizations have the obligation to retain seven years of backup in some form. The more data you keep online (data that's not archived or purged), the more resource consumption it takes to perform any operation, leading to more I/O (input/output), more compute (servers), and more operational costs.

Can we archive any of this data? The answer should be yes because it's impacting performance, maintenance costs, etc. But that's not what's considered. The business does not understand the true cost of ownership or the impact of their decision to keep the data, so they retain it all, just in case. This goes back to not providing the right supporting information when we are asked if we can archive or purge the data.

What if, instead, we posed this question to the business about archiving data: *Can we archive any data older than eighteen months, as it is costing the company $5,000 per year to retain the data in the table?* If they know the total cost of the data and you pass the cost onto the business, I bet their answer would change.

However, the average tech person does not know the cost or impact of their decision to keep the data. They're running maintenance on this process every day. For every gigabyte of data they have, there's a cost. And there's also a code cost that increases as the amount of data increases. So it is important to understand the relationship between the cost of code and the cost of data, as growth in one impacts the other. Just as for code and applications, we need to ask questions about data retention and data life cycle and put a process together based on what it's costing us on average—this is your real cost of data.

Remember, if I tune a query and show resource savings for one execution versus one year, the one-year savings will tell a more impactful story. This is the mindset that needs to prevail.

FINOPS SHOWS PROMISE FOR GAINING FINANCIAL TRANSPARENCY IN TECHNOLOGY

One of the biggest shifts to occur with the move to the cloud is how we pay for technology and applications. The industry has moved from all-you-can-process on a server for on-premise data centers to a variable or utility compute model. According to Apptio, in their report *FinOps: A New Approach to Cloud Financial Management*, this means that "micro-optimizations can happen at the team level each and every day to change the shape of the cloud spend … It's a world of OpEx [operational expenses] instead of CapEx [capital expenses], completely changing how finance is reported and managed."[60]

As a result the traditional procurement model for expenses has been upended, putting the spending power in the hands of engineers who are managing these applications and infrastructure with very little regard for what it's costing the company in operational expenses. Apptio describes the grim reality of this situation as "engineers making financial commitments to the cloud that affect the bottom line of their companies while finance teams struggle to keep up with the pace and granularity of spend."

Apptio further goes on to highlight just how everyone loses when there is no transparency into the costs of cloud storage:

- Engineering spends more than it needs to with little understanding of cost efficiency.
- Finance teams struggle to understand—and keep up with—what is being spent on a mind-boggling number of options. (AWS alone has approximately 300,000 SKUs and additional thousands of new features per year.)
- Leadership doesn't have enough input into how much will be spent or the ability to influence priorities.
- Procurement isn't a deliberate participant in its own outsourcing.

Hence, the discipline of FinOps has emerged to save the day. Apptio states, "FinOps brings together finance, technology, and business leadership to master the unit economics of

---

60   451 Research, "Cost Management."

cloud … With FinOps, each service team or product owner has the data that enables them to have some control over their spend and to make intelligent decisions that ultimately impact the cloud bill." In short the communication and collaboration between finance and technology that I talked about in chapter 4 is already occurring for companies that have developed a FinOps model to manage technology spend.

We need transparency into the true costs of applications, code, and data to understand the true costs of our systems. This can only occur by forging and strengthening partnerships between technology and the CFO's office.

CIOs who have successfully optimized IT costs have teamed with other C-suite executives, particularly the CFO, to collaborate in enterprise cost optimization initiatives.

—IRMA FABULAR, SENIOR DIRECTOR ANALYST, GARTNER

Gartner lists "IT financial transparency practices" as one of the ten techniques for IT cost optimization to navigate budget constraints in a volatile business environment and offers this advice to IT leaders:[61]

Understand how IT services are being delivered and what the associated costs are for IT operations. Work with CFOs and finance personnel to associate general ledger entries (asset-based view) to technical costs and related costs for business services.

---

61 Laurence Goasduff, "Ten IT Cost Optimization Techniques for Private and Public Organizations," Gartner, March 25, 2021, https://www.gartner.com/smarterwithgartner/10-it-cost-optimization-techniques-for-private-and-public-sector-organizations.

Such transparency in enterprise IT spending enables you to deliver business value and optimize costs via IT services.

While a giant step forward in terms of progress, FinOps is a mostly a nice theory that has yet to be fully embraced or articulated. FinOps does not consider all the different cost drivers, such as inefficient code, and that the current tools and services in the market do not fully leverage operational data that would allow FinOps teams to identify the root causes of increased technology costs. Instead, they stop at the resource level (servers and devices) and do not go deeper into the code level. This will be the next frontier, as there are diminishing returns with the current tools that fall short of actions like rightsizing, powering down or off nightly, and "reserving," or prepaying, for your server for one to three years out.

The industry has yet to instrument techniques for micromanaging application and infrastructure changes on a daily basis in order to fully realize the benefits of a FinOps collaboration. Still, two great resources for any company looking to establish a FinOps discipline are as follows:

- *FinOps: A New Approach to Cloud Financial Management*, Apptio
- *Cloud FinOps: Collaborative, Real-Time Cloud Financial Management*, O'Reilly Media

WHAT IS YOUR TOTAL COST OF OWNERSHIP OF TECHNOLOGY?

You wouldn't buy a car based on the list price alone. You know there are other things that go into that purchase, such as gas, maintenance, and depreciation. You would factor all these things in, and more, before deciding which car to purchase because you want to know the Total Cost of Ownership of that car. Why aren't we doing this for technology systems?

If we don't start to adopt this mindset now, the cost of inefficiency and, more importantly, the cost of inefficiencies over time will become overwhelming. Just as a more concerted approach to combating climate change should have been implemented decades ago, my fear is that if we let things continue as they are, the millions of dollars in unforeseen expenses to all businesses trying to compete in today's 24/7, always-on world will become trillions. Aside from the impact on corporate bottom lines, no one wants to see a world in which the costs of technology will outweigh its benefits.

When purchasing an application to provide a function for a business, many will compare at least three vendors on the basics, such as functionality and support. But a more detailed analysis of the Total Cost of Ownership of that application over three years might be a better approach because if two apps are essentially comparable (do the same thing, have great support, similar purchase price), this will distinguish the best choice. Another way to look at it is this: *Which application will cost me the least to implement, manage, and maintain over three to five years based on my business model and growth metrics?*

Tools exist to help figure this out, but they fall short because they do not have actual metrics from different companies. Often they are using the vendor's estimates and projections, so buyer beware. Let's look at some models from cars to the cloud, and you'll see some trends evident that we could apply to the TCO for your technology environment, servers, and apps.

TOTAL COST OF OWNERSHIP FOR AUTOMOBILES

When you're looking to buy a car, you are unlikely to consider solely the list price when making a decision. One measure you might also consider is the Total Cost of Ownership for three years, or five years, because, if you're like most people, you're planning to drive your new car for a while. (Why do we not think this way about technology? More about this later.)

Edmunds, a trusted online resource for automotive inventory and information, offers an online calculator (https://edmunds.com/tco.html) for estimating the total five-year ownership for cars. The inputs Edmunds uses to make these calculations include car depreciation, interest on financing, taxes and fees, insurance premiums, fuel, maintenance, repairs, and any federal tax credit that may be available.[62] There is a lot that goes into the cost of buying a new vehicle beyond the cost of the car itself. Edmund's tool goes so far as to estimate the cost of each of these variables each year for five years to determine the true cost of ownership. These benchmarks can be used to compare different vehicles under consideration using the same methodology to help customers make an informed decision on which car may be the most cost-efficient over time.

TOTAL COST OF OWNERSHIP FOR APPLICATIONS

In their information technology glossary, Gartner defines Total Cost of Ownership as "a comprehensive assessment of information technology (IT) or other costs across enterprise boundaries over time. For IT, TCO includes hardware and software acquisition, management and

---

62   Edmunds, accessed November 26, 2022, https://www.edmunds.com/tco.html.

support, communications, end-user expenses and the opportunity cost of downtime, training and other productivity losses."[63]

How can the total cost of software be calculated? An article in *CIO* magazine, titled "Calculating the Total Cost of Ownership for Enterprise Software," reported: [64]

The Total Cost of Ownership (TCO) ... is a critical part of the ROI calculation. However, it is often ignored or woefully underestimated.

According to the article, the costs to help estimate a realistic TCO over the lifetime of the software when comparing cloud, custom, and off-the-shelf software include the following:

- Start-up costs (including software and hardware costs, implementation, data migration, user licenses, and more)
- Operational costs (including maintenance and support, data center, depreciation, patches, user licenses, training, enhancements, and more)
- Retirement costs (including data export, keeping archived systems, and inactive licenses to preserve audit trails)

Note each category includes varying licensing costs.

Software Advice, a Gartner company that provides advisory services, research, and user reviews on software applications for businesses, offers a Total Cost of Ownership calculator (https://softwareadvice.com/tco/) that analyzes the TCO for on-premise software systems compared to software-as-a-service (SaaS) systems.[65]

---

63 "Glossary," Gartner, accessed November 26, 2022, https://www.gartner.com/en/information-technology/glossary/total-cost-of-ownership-tco.

64 Chris Doig, "Calculating the Total Cost of Ownership for Enterprise Software," *CIO*, November 19, 2015, https://www.cio.com/article/242681/calculating-the-total-cost-of-ownership-for-enterprise-software.html.

65 Software Advice, accessed November 26, 2022, https://www.softwareadvice.com/tco/.

The categories they use to make these comparisons include the following:

- Licenses and subscriptions (those licenses again!)
- Installation and setup
- Customization and integration
- Data migration (from on-premise to the SaaS platform)
- Training
- Maintenance and support
- Hardware

This tool from Software Advice projects costs over ten years. Interestingly, while the cost of SaaS is considerably cheaper at the onset, the total cost estimates can be nearly identical when you factor in all the additional costs over time.

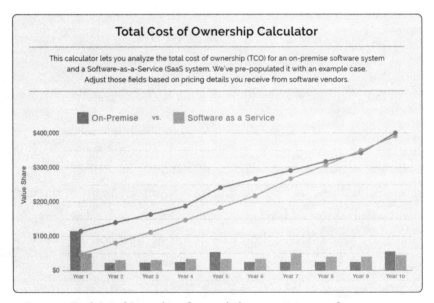

Figure 12: Total Cost of Ownership software calculator, on-premise vs. software as a service. Source: Software Advice, a Gartner company, https://www.softwareadvice.com/tco/

My experience with clients at Fortified looks more like this scenario:

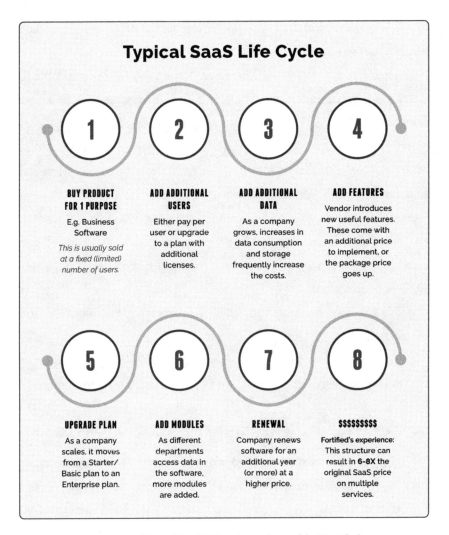

Figure 13: Typical SaaS life cycle as observed by Fortified.

A company buys a SaaS product for one purpose—to solve a problem identified in the business plan, whether that's organizing a customer database, improving accounting systems, or streamlining workforce management. Typically, over the life of a SaaS product, a company will add users, data, and features offered by the provider to get more out of it. This is commonly known as "scale," and any company that wants to remain in business needs to do it. So they will upgrade from a "basic" plan to an "enterprise" plan to accommodate business growth.

They will add more modules (i.e., make the solution available to different departments within the company). This, of course, will raise the cost of the SaaS product as the company grows. We've seen clients realize increases up to eight times the original price of a SaaS product as they move to multiple services due to the way SaaS providers sell to customers and then engage and upsell them over time. Sometimes companies will choose to migrate to a different SaaS vendor because the company's needs have outgrown what the SaaS product can provide, adding more start-up costs to the equation, and the whole process begins again.

No one is thinking about these future expenses; they're all too concerned with getting the job done. They'll pay up front to make that happen, and as long as it's doing what it's supposed to do, the compounded costs are not front and center. This needs to change.

TOTAL COST OF OWNERSHIP FOR AZURE WORKLOADS

To help potential customers who are thinking of migrating their on-premise workloads to cloud-based Microsoft Azure, Azure offers a tool that recommends which of their services can help the customer save money if they were to migrate. Not surprisingly this tool (https:// azure.microsoft.com/en-us/pricing/tco/calculator/) is far more technically detailed than the other examples above.

The categories they offer to define workloads are as follows:

- Server workload
- Database infrastructure
- Storage infrastructure
- Networking bandwidth

Within each of these categories, there are several fields to enter more details to understand each of the cost inputs for that category. For example, within the server category, further input as to the type of on-premise workload and server environment, operating system, operating system license, number of servers, and more is requested.

Using a series of industry averages, the tool then makes certain assumptions that can be dialed up or down by the prospective customer. The resulting report provides the estimated cost savings for moving from an on-premise solution to Microsoft Azure over a five-year period. No yearly breakdown of costs is offered in this example.

While these inputs are valid for companies considering a cloud migration, they don't provide a complete picture. When organizations have thousands of servers, 10 to 20 percent YOY data growth, increasing transactional requirements, and continually changing applications because of new application releases, not to mention application or workload inefficiencies, how can you really know the cost of the cloud option *today*? Or *five years* from now?

It's a hard question to answer, but based on our experience, many companies are being told one number via these simple cloud calculators, but once they migrate, their cloud bill does not reflect the sticker price. In addition to asking what the cost is based on all these inputs, a company should be asking, *What is the cost of data, code, and resources that will be deployed in the next year?* Now we know most price estimates will be off by a certain percentage. But how much leeway is your business comfortable with?

Let me share two cloud migration experiences where two different organizations migrated one of their mission-critical applications to the cloud. The first company wanted to vacate their data center by a certain date and save money. Their existing hosting costs totaled $3 million per year. Because of the deadline, the client migrated their applications to the cloud as is. Once they migrated their application with all the supporting infrastructure, their cloud bill was $350,000/month or $4.2 million/year—more than the total cost of supporting all their applications in the data center.

The second migration was driven by a partnership with one of the big three cloud providers, as both companies received some publicity and benefited from each other's services. It was not cost-driven, but after this client migrated to the cloud, they had some performance issues because the cloud is not built to handle their type of workload the same way they managed it on-premise. So the client had to allocate additional resources to enable the application to perform well enough to support the business. Now the client is spending $1.5 million/month or $18 million/year to support the application, which, if housed on-premise, would cost one-third that amount. These numbers do not factor in licensing, development, or support costs, as these are in addition to the monthly costs as noted earlier.

ESTIMATING THE TCO FOR YOUR TECHNOLOGY ENVIRONMENT

It's not common practice in the industry to calculate the TCO for your technology environment or the hundreds of applications your team is supporting. But wouldn't it be nice to know the real cost of supporting each application? Just as my client shared in the opening of this chapter, most of us have purchased an application from a vendor who recommended the hardware to support the application for three years, but at the end of year one, the servers were at capacity or had performance issues, and additional resources had to be added. So now the TCO just increased, and your budget needs to change to capture the recent growth while factoring in the new three-year TCO and ensuring the applications continue to meet the performance SLAs of your customers.

Why are the vendors getting their estimates and requirements to run and support the software so wrong? Because while the vendors are good at building software to solve a business problem, *they are not experts on how you will use the application to run your business.* This list highlights some things vendors do not typically know about your business and why their estimates vary so much:

- Type of infrastructure
- Architecture (high availability, disaster recovery, etc.)
- Type of business data
- Transactional and data volumes
- Costs (licensing, hardware, etc.)
- Processes and policies
- Regulatory or industry requirements
- Unique workload requirements (reporting not on the vendor database)

Today most organizations leverage software, processes, and metadata (virtual tagging assets) to manage their technology assets, capacity, and utilization, but they are focused on the needs of today. Some capacity planning tools allow you to forecast capacity at the server level, but much of this information is isolated in separate systems across the enterprise. This makes it difficult to capture and merge the capacity data with all the other data required to determine the TCO for the applications.

A NOVEL APPROACH FOR ESTIMATING TECHNOLOGY INVESTMENTS

How can we capture the TCO of technology investments and applications today and for the next three to five years so we can accurately budget, plan for resource needs, and meet business expectations for growth?

Consider the following:

Initial Costs + Ongoing Costs - Cost Takeouts = TCO for Software/Applications

where

Initial costs =

- Installation and setup
- Customization and integration
- Data migration (to and from the SaaS platform)
- Training
- Perpetual license

Ongoing costs =

- Maintenance and support
- Licenses and subscriptions (SaaS)
- Hardware (compute, memory, storage, bandwidth)

Cost takeouts =

- Rightsize systems
- Optimize costly processes
- Shift schedules from nonpeak times (workload rebalancing)

- Downgrade to nonpremium editions or subscriptions
- Implement data life cycle policies

As the system grows, it will increase the hardware resource requirements, such as compute, storage, memory, and bandwidth.

There are two important things to note in this approach that differentiate it from conventional thinking in the marketplace today.

1. Today's models don't factor in the growth rates for data, business transactions, and the resource consumption for each application. As today's systems and businesses grow at higher rates, it becomes more important to factor this growth into the TCO model to accurately capture the true spending that's required.

2. While we have been focused on spending, another important variable that often goes overlooked in the TCO model is where there are opportunities to realize a cost savings (i.e., proactive cost takeouts). Within any system we've reviewed at Fortified, we've always found ways to reduce costs within an application deployment. As we stated in previous chapters, we have been in the abundance mindset, and because of this, models are focused on increasing resources as business as usual, but it doesn't have to be this way. This is data that every executive leader wants, as they get paid to save money and bring more value to the business.

3. Look at it another way; there is a lot of money going to Microsoft, AWS, or Google to sustain these inefficient systems. Would you rather be feeding the market caps of these companies or investing in your own company's future?

THE ONE COST NO ONE'S CONSIDERING: THE COST OF GROWTH

In my line of work, CFOs are often surprised (and disappointed) when their systems run out of capacity, or they have to buy more storage, CPU, memory, or bandwidth to keep the business going—costs that weren't in the budget. CFOs feel helpless because if they don't approve the expense, the system will not operate. If we can understand the inefficiencies or potential cost takeouts currently driving these extra cost outlays, we can prevent them from happening. Your licensing costs may be fixed; you can estimate that. Your growth rate is not fixed, but you can estimate that as well since it's most likely a KPI the sales team will be measured against.

Estimating the savings you will realize if you optimize noneffi-cient code before it is promoted to production (at best) or causes an outage (at worst) takes a bit more planning and insight. But it's necessary if we want to keep up with the current rate of growth busi-nesses are experiencing while balancing costs.

Not convinced that one inefficient piece of code can disrupt the entire system or lead to millions of dollars in unanticipated costs? Well, if yours is like most technology environments, an application can be installed, and then nobody looks at it for ten or twenty years. Remember when you learned about compound interest when you opened your first savings account? Well, the cost of that inefficient code compounds as well, and if you're looking at decades, this com-pounded cost of inefficiency adds up.

What is the value of a cost takeout? Unfortunately this is usually realized too late—after the fix is made and the cost incurred. Financial modeling and analysis lets you see how much can be saved (i.e., not spent) if the system is operating efficiently. Your company is likely bearing costs that you think are the cost of doing business, but they don't have to be.

This is timely information because I can't believe how fast companies are readjusting their budgets and freezing spending due to the economic environment. But data is always growing, and technology is critical to every organization, as they accelerate their digital transformation. Everyone must keep spending more money on technology to compete and excel.

Another challenge is when the business needs to reduce spending. What do you cut in technology that will not impact security, availability, performance, and innovation? You're going to have to find some cost savings, but where are the opportunities, and how much will they add up to?

Your company is likely bearing costs that you think are the cost of doing business, but they don't have to be.

Now we optimize and forget it. We need to prioritize optimizations over time—at least quarterly, semiannually, or yearly—if we don't want these cost savings to go unrealized. Remember, each period between optimizations results in more unintended costs that could have been saved.

The truth is it's hard to know what your cost takeouts will be. We don't know where, when, and what the costs are, but could we? We have the data for nonefficient code, transactional metrics, and cost of technology (cloud bills). Now we need to merge this data with some assumptions on percentage optimization so we can start to predict our total cost savings.

It is my hope that having the information presented in this chapter will help CDOs, CIOs, CTOs, and even CFOs understand the importance of tracking and managing system costs versus going on a hope and a prayer that the status quo will prevail. Every company will have its own technology budgets with gates and guardrails. You want to make sure you're protected in case of an unforeseen emergency. You want to have enough capacity, be secure, and protect your data against the worst possible event, or your system, and your business, could be destroyed.

But it's been our experience that the degree of inefficiency is so great that even cutting it by 50 percent could still leave enough cushion to weather anything this unhinged world can throw at us. And the magnitude of cost takeouts that are not being realized is causing significant profitability drains on your business.

If you don't keep your car's engine tuned, its efficiency will go down, and it won't be there for you ten years down the line. If you do, it will need less maintenance and burn less fuel. Don't you want the same for your technology?

There are two paths you can take. You can go down the traditional path and be told these are the resources it's going to take, and you keep increasing your technology spending. And if you don't, there will be outages and negative performance for the users.

Or you can go down this new path that takes a lot fewer resources, reduces your costs, and allows you to increase your margins. And the difference between these two paths over time—that shaded area between what's below the efficiency line and what's above the inefficiency line—that's your savings.

A CFO'S PERSPECTIVE ON FINANCIAL TRANSPARENCY INTO TECHNOLOGY SYSTEMS

Allen Parker is the CFO of Zillow Group, where he oversees the real estate software's accounting, finance, legal, M&A, treasury, shared services, and strategy functions. Prior to Zillow, Allen served as VP of finance in various roles at Amazon for thirteen years, where he helped architect and scale key finance structures and processes for Amazon devices, operations, e-books, Amazon Pay, and more to support rapid and robust growth. Zillow and Amazon both know a lot about scaling and efficiency. So I sat down with Allen to see what insights he could bring to our discussion of financial transparency.

Ben DeBow: As a CFO, how do you ensure a productive relationship between your technology and finance teams?

Allen Parker: Our conversations start as early in the process as possible. Whenever there's a new product or service or initiative, we want to be involved up front in the business requirements doc so that, as they build the product, the technology requirements flow all the way through to our accounting systems and journal entries. We rely on technology general controls to understand where the source data is coming from, how that source data is protected, and how it eventually gets reflected in the financial statements, which means much fewer manual journal entries and calculations.

Technology plays a very big role in the success of my team to support the business as it scales, whether that's interacting with accounts payable, vendors, or managing receivables. If you were scaling fast and you did not have a good connection between your ERP [enterprise resource planning], accounting, and other business systems, you would have to bring in more people to complete more manual entries to support the business. Having technology involved up front means you need a smaller number of higher-judgment people who are inspecting the inputs and outputs versus processing reports, creating Excel sheets, and booking journal entries.

Another area where we benefit from technology is financial planning and analysis. The more we use the data we have on our customers, business processes, and systems, the more we can create algorithms, forecasting, and analysis models. We can be a lot smarter and more efficient in trying to predict what current history may mean for our future progress. We can do a lot more analysis with more granularity. Technology is a huge component of success not only for the business but also in my organization's ability to support the business at scale with growing complexity.

Ben: How can you ensure financial transparency into your technology metrics?

Allen: When you can reduce manual entries and bring all your information together into one single source of truth that the business and finance analytics teams are looking at, you're always on the same page. Data is always refreshed in the same way for access by anyone who needs it. You can't have different people working in different spreadsheets across the company; you'll never get financial transparency that way.

When you manage revenue, by the time you see your final numbers at the close of a period, you're probably already two months too late in responding to some of the issues there might be. If I was really focused on driving revenue or whatever my output metrics are, I really should be managing the inputs (industry trends, customer trends, buying behavior) to better project what those outputs will be.

I'm managing the inputs within a certain set of control limits, and understanding when they're moving outside of those control limits, I can react to business signals that will eventually affect my revenue with very little latency. In other words as soon as I get the signal, I know how it's going to impact my numbers, and I don't have to be surprised when they finally come in.

You need a strong technology platform that allows you to get to the data when you are managing several diverse product lines to understand where to allocate future investments. You need to be sure you understand the data at a level of transparency and granularity as the business grows more complex. A lot of new businesses experiencing hypergrowth are growing in both scale and complexity. As a CFO it's impossible to get your arms around those businesses and really expect to be serving them in a way that influences decisions without a really good understanding of core data. You need the systems in place before the businesses have scaled; otherwise, you'll spend all your time backtracking.

Ben: Amazon is probably one of the most efficient companies out there, or at least it looks that way when I order my batteries on Friday at 3:00 p.m., and they're on my doorstep Saturday at 11:00 a.m. Are there any lessons you've learned about technology efficiency working at Amazon?

Allen: Jeff Bezos had a process that worked, and it was very transferable across a lot of different businesses—if it's knowable, you should know it. You not only have to improve (i.e., your revenue should go up) but you also need to know you're improving. If revenue goes up but it's due to something completely outside your control, that may be great for a period of time, but it doesn't mean you've improved anything. At some point that variable that was outside of your control will change, and then you're going to be surprised when, suddenly, revenue isn't as good as you want it to be.

Ben: How do you apply the Total Cost of Ownership to your technology investments?

Allen: Cost efficiency means investing in something that will allow me to reduce the cost burden not just for today but also at scale. If there's a big initiative and I'm trying to drive an ROI calculation, I want to ensure I understand holistically what all the costs are versus just assuming the direct costs without appreciating them over time. It goes back to decreasing manual effort, as this has a cost. If I drove an initiative and we weren't building it right and it required a bunch of manual people to support it, I should be including that cost into the calculation.

Something may not cost me a lot today, but if I was scaling at 10x, it could get extremely expensive. Cost efficiency means looking at those investments that allow me to reduce my AWS spend or eliminate other corporate applications. (Managing other corporate applications requires more technology support and more fixed costs without the benefit of one platform or "source of truth.")

Costs are like an iceberg. You can see the ones that float on top. But you don't see the 75 percent down under the waterline—these are still real costs that will permeate throughout the business as inefficiencies or wasted time. If you don't pay attention to it, the iceberg will start to rise.

—Allen Parker

If you take nothing else away from this chapter, know this: there are more excess costs being spent now than you think. And it's only going to get worse as data and businesses grow. What's it worth to you to assess your systems now? (You can find an assessment to help with this at my website, https://www.bendebow.com/assessment.)

As you can see, there are many variables that impact the TCO for applications. How do we instrument the systems today to capture this important data so that CFOs and CIOs have an accurate understanding of TCO and the three-year spend while meeting business goals and objectives today? We need to start by gathering the right data about each application and integrating the data so we can instrument the TCO analysis within your organization. It's ideal to have this data at the beginning of the project, but businesses need near-real-time financial views on the TCO of their application deployments at any stage of their life cycle, especially if they'll be with you for the next twenty years.

In their article "Nine Winning Actions to Take as Recession Threatens," Gartner advises:[66]

Using a shared framework to evaluate cost initiatives ensures that resources are traded and cut for the greater good of the business—in this case, to capture efficiencies while preserving funding for critical digital initiatives. Strategic cost optimization that moves conversations from cost to value helps ensure resource decisions are strategic, not tactical.

Rightsizing capacity and removing inefficiencies are real opportunities that should be proactively calibrated to shift the conversation from cost to value as Gartner advises. Measuring these impacts using financial data would create tremendous levels of confidence and transparency between the C-suite and tech teams. Getting a server healthy removes costs today and in the future not only by increasing the ROI of your tech spend but also by directly impacting the financial profile of your company. It's not just the up-front costs but also reducing the Total Cost of Ownership over time that will create real value for your company.

---

66 Sanil Solanki and Alexander Bent, "Nine Winning Actions to Take as Recession Looms," Gartner, November 26, 2022, https://www.gartner.com/en/articles/9-winning-actions-to-take-as-recession-threatens.

KEY TAKEAWAYS

- To optimize the costs of technology, you need to understand not only the costs you're incurring today but also how these costs will compound over time.

- System efficiencies can be identified by rightsizing capacity as well as looking at applications, code, and data to find areas of optimization.

- The concept of Total Cost of Ownership used in pricing cars can be applied to software and servers as well. We need to build these types of tools in the technology industry to truly understand system waste and inefficiencies.

- The issue is not one of simply cutting costs but rather proactively taking costs out of your systems in the first place. This means identifying cost takeouts to minimize waste and optimize resource consumption.

- Gaining financial transparency into technology systems requires collaboration between technology and finance from the outset of any new project and clear lines of communication thereafter.

CHAPTER 7

Actions for Improving Technology Health to Drive Business Value

I f you've made it this far into this book, I hope you're thinking about the efficiency of your technology systems, as well as the long-term costs to operate them. In this chapter we're going to consider the actions you can take to make significant improvements to the way your systems run that will have resounding effects throughout your organization.

To kick things off, consider this story about an actual Fortified client in the global hospitality industry. The company had migrated one of their mission-critical applications to a leading cloud platform. After migrating they experienced significant performance and stability issues as do many of the clients we serve. They engaged Fortified because of our expertise in identifying and solving problems.

We identify the core issues impacting the business or technology group that are keeping them from meeting their goals and objectives. There's a difference between issues that are impacting the business and those that are affecting the technology teams. One of the challenges organizations have is that there's always something you can change, but just making that change is not always enough to solve the problem. At Fortified we don't identify just issues and how to solve them but whom they impact, and then *we map those issues to actions.*

There is an abundance of monitoring, performance, and operational data about every system, but rather than dive into the charts and tools willy-nilly, I encourage my team to take a step back and answer this question: *What are the one or two actions we can make on this system today that will have the biggest impact tomorrow?*

Within a day of being engaged by the hospitality company, we identified the most critical issues and prioritized the actions that would have the greatest impact on improving the performance and stability of the system with the least amount of effort and risk. One of these actions was to optimize a critical process that took around twenty-four hours to run, which impacted the financial health of the company. After the optimization, which took less than one day to complete, the process ran in twenty minutes and allowed the company to solve their biggest issue. They were amazed, but still, the CFO questioned why our hourly rate was "so high." I maintain this is the wrong question. Rather, the question CFOs should be asking is, *What is the value to your company for fixing critical business processes in a timely manner?*

Over the next two months, we optimized about forty more processes, and each time we did, we captured the impact of the tuning effort before and after our work. This allowed our team to show the financial savings (i.e., business value) of the services we provided. We provided technical savings for the technology teams, financial savings for the CFO, and time saved for users, leading to reduced wait times and an improved experience. For this CFO, we translated the reduced CPU cycles, reduction in storage throughput, and processing time into terms the business understands—dollars and cents.

A client with an issue needs the issue solved. Every time we work with a new client that has the goal of saving money, we set out to identify the most inefficient business processes that are impacting the company's health and performance. With every client, we look to identify *the ten most inefficient systems or processes that are impacting workload health.*

One of the most frequently asked questions in the application development world and especially in the database space is, *What are the ten most inefficient, least optimized statements so we can tune them?* The challenge with this question is that most database administrators have a different interpretation of what is bad or good and look at the statements independent of time and financial value and/or only look at the statements at the server level. Ideally we need to review those statements across the enterprise that are costing the company the most money and are impacting application SLAs. We need to change the way people approach this basic question and help application developers better understand the impact or value of them optimizing the code or process.

WHAT DOES IT MEAN TO OPTIMIZE BUSINESS PROCESSES?

Optimizing (or refactoring) processes, reports, and applications to address inefficiencies may include the following:

- Configuration changes to the OS (operating system), platform, or application
- Optimizing the server size or service offering
- Auditing and disabling legacy processes that are no longer required
- Changing the timing of a process
- Changing or reducing the ongoing maintenance of a process
- Optimizing the indexing strategy
- Archiving and purging data

- Optimizing the code

- Refactoring data structures

These recommendations are prioritized in the order in which we recommend approaching problems based on the balance between risk and effort (i.e., low effort/low risk and high effort/high risk vs. impact).

	LOW RISK	HIGH RISK
HIGH EFFORT	Changing or reducing the ongoing maintenance of a process Optimizing the indexing strategy	Archiving and purging data Optimizing the code Refactoring data structures
LOW EFFORT	Configuration changes to the OS (Operating System), platform, or application Optimizing the server size or service offering	Auditing and disabling legacy processes which are no longer required Changing the timing of a process

Figure 14: Optimizing business processes based on evaluating risk versus effort.

Any change to the infrastructure, applications, or database that will help make the system more stable, reliable, or efficient needs to be measured both before and after the change is made in order to quantify its value.

In the first chapter of this book, I talked about a better way to measure technology; in fact, it was the name of the chapter. Your perseverance in reading has paid off because now I'm going to go into detail about this better way—or *new way*—compared to *the old way* of doing things.

THERE IS A BETTER WAY TO MEASURE WORKLOAD HEALTH

We would stay with my example of the hospitality client. The old way of solving their issues would have been as follows:

The developer, in their technical mindset, is ecstatic because they tuned a report and reduced the run time and resource consumption. After tuning the report, they send an email to the key stakeholders of the company with the following information:

Or if there are multiple reports, the email might include a chart with the following information:

Object Name	Resource Impact		
	CPU Impact	IO Impact	Duration
report_by_industry	70%	65%	75%
report_by_type	75%	78%	90%
report_by_geo	73%	65%	80%
report_by_rate	78%	79%	99%
report_by_material	55%	85%	86%
report_by_cost	65%	75%	99%
report_by_margin	75%	75%	85%

Figure 15: Typical optimization results distributed internally.

This might look great to the tech people reading this, but what does it mean to the business folks?

* Do they even know what I/O is? (Input/Output = Storage)
* What does 65 percent faster actually mean?
* How much money did this save them?
* How many times is this statement executed in a day or year?
* How much do they save each time it is run?
* What is the ROI on their spending to hire us (Fortified) to figure this out?

These questions point to the difference between the old and new ways of assessing server health and efficiency. Simply put, without these results, the findings have no context in terms of business value or impact.

CHANGING TECHNOLOGY WINS INTO BUSINESS WINS

Today people are communicating wins in technology terms to a business audience that doesn't understand technology. Not only do we want to determine the impact of these changes in terms of *here's what you had before, and here's what you have now*, but we also need to change the language we're speaking to the language of the business.

We also need to change how we ask the business to approve budget and time for people to work on the ten worst-performing (least optimized) database statements. Today database administrators may send the ten most problematic database statements to developers as a task or report without much explanation as to how working on these will improve the system. Because we are not clearly explaining or presenting the potential impact to the business, these changes do not get prioritized, and therefore, it is system status quo. Therefore, we need to change the language not only in terms of the ROI/impact but also when we are assigning or proposing optimizations to developers and product owners. Since the product owners own the feature road map and

> **We need to change the language not only in terms of the ROI/impact but also when we are assigning or proposing optimizations to developers and product owners.**

budget, ultimately we need their buy-in and support to get optimizations into the application road map. By doing so, we can gauge the net impact to the system, beginning with when the change was implemented, so they can plan accordingly as they set timelines.

So here's what we did for this client (a.k.a. the new way of doing things) to communicate our value in terms they'd understand: At the start of every engagement, we baseline the system from a performance, run time, and metadata perspective. This allows the team to go back and show the value and impact of the changes at any time.

For the technical audience, which understands the resource cost of CPU, I/O, and duration, the results report might look like this, but typically they will speak in terms of one execution instead of the one- or three-year cost or impact.

Object Name	1 Year Savings		
	CPU	Reads	Duration
report_by_industry	84,663,502	250,152,130	135,018,154
report_by_type	260,144,625	75,044,213,160	498,617,339
report_by_geo	1,255,144	2,674,264	2,097,436
report_by_rate	591,070,324	39,475,666,789	722,294,565
report_by_material	24,939,994	818,238,750	66,560,298
report_by_cost	175,719,851	16,260,673,350	703,210,226
report_by_margin	1,019,346,998	114,687,195,128	4,580,345,160
Totals	2,157,140,437	246,538,813,570	6,708,143,178

Figure 16: Optimization results with an eye toward future impact.

We need to convey the larger story because the code is not running for one day and then stopping. This has real impact. *So for the business,* we use a different language to present the information, keeping it in terms of annual impact to better demonstrate the true value:

- Total report duration was reduced from 127,000 minutes to 15,000 minutes for the 57,875 report executions. (This is a reduction from 35 days to 4 days.)
- Average report execution was reduced from 2.19 hours to 26 seconds, per report.
- Total estimated annual cost savings for technology is $31,807 in cloud hosting costs, representing a five-year savings of $159,035.

No one likes to see the hourglass running while a report is trying to execute. It's a waste of everyone's time even if it's just a few seconds. Put another way,

- these reports used to run 127,000 minutes or ~35 days over a year;
- if we were to annualize our tuned reports, we'd reduce that time by 112,000 minutes, or 31 days; and
- we gave 31 days of "waiting" back to the business by saving their users valuable time.

This, I propose, is the new way of presenting information on workload health and technology value.

With this approach, the business can see the impact that previously only the tech team would understand. The change didn't just have a one-day impact but it will also continue to impact the systems over many years to come. And if it's an application that will remain in use for twenty years, you've just created some real value for the business that they will see and understand. In our experience at Fortified, this can amount to millions of dollars in savings that can be reinvested elsewhere in the business for further improvements or business gain.

Today these calculations are all done manually, even if a developer did choose to show the before and after pictures. In the new way of doing things, we need to build tools and processes to engineer this type of mindset so that the business can focus on measuring the value of their employees versus how many widgets they produce or how many tickets they log. What's more, if the business could learn to do it for themselves as often as they needed, it would justify the expense of companies like Fortified to be able to create and compound these savings with every execution.

This is the level we need to aspire to if we're going to change the prevailing mindset. Take action and then measure the impact of the action. If we can systematize both, we'd be making real progress. Admittedly it's easier to systematize measurement of the actions; we'd still need a good deal of human interaction to understand what changes need to be made in the first place.

In summary, at a high level, to change technology wins into business wins, we need to

1. change the terms we use to be relatable to the business and senior management;

2. focus on the one-, three-, and five-year impact of each change; and

3. build the tools and processes to help businesses recognize the savings and efficiencies within the technology ecosystem as business needs dictate.

WHO IS COMMUNICATING TECHNOLOGY'S WINS?

Another factor is having the right people in place to determine and communicate the findings. Very few people ever show clients the exact changes that need to be made. Additionally, once those changes are made and the findings presented, nobody talks about the net impact of the changes over time.

Think of this experience at the gym: A trainer greets you, looks at your intake form, and says, "You're not healthy. You're not where you want to be." You've identified all these things you want to fix, or goals. So the trainer creates a plan for you to follow. You're going to do these three things every morning and these two things three times a week.

You start following your plan. Then you see the trainer, and they measure your progress with your Fitbit, Apple Watch, or other application and say, "Your stats are improving. Your weight is down three pounds, inches are coming off, BMI is down. Congratulations!"

We don't express technology in terms of business results today. There's no designated "trainer" telling clients the real-time capacity they saved since the code was optimized in the last application release. We're not measuring impact or value of the changes effectively. We're not telling clients the actions they should take, and we're definitely not measuring after any changes are made. It's not just making the technology changes but also proving and recognizing their business value that is missing from the way things are done today.

> **It's not just making the technology changes but also proving and recognizing their business value that is missing from the way things are done today.**

Based on the workload health metrics we've discussed thus far, what changes can you make to system processes, and then how can you measure the impact of those changes and demonstrate value back to the organization as a result? To help answer this question, let's first consider the notion of technology debt.

WHY DO WE HAVE TECH DEBT?

As I've mentioned frequently throughout this book, most application development is centered on new features that need to be developed quickly because the market cannot wait any longer. This is understandable because customers want newer, bigger, and better (or, as the case with iPhones, smaller, then bigger), and any business that wants to remain in business builds things to keep customers happy.

But this mindset has led to the notion of technology debt (a.k.a. technical debt or tech debt), which results from the frequent deployment of "quick and dirty" solutions at the expense of a well-thought-out approach. Tech debt results in all the things you would love to do to enhance your system but haven't gotten around to doing. Often this includes fixing or improving existing technology, and it falls by the wayside in favor of newer, bigger, and better. It's like the old adage "Don't put off until tomorrow what you can do today." The definition of *tech debt* is basically putting everything off until tomorrow. And of course, tomorrow never comes in terms of tech debt because once the code is deployed to production, it's not getting optimized or changed unless it breaks the system.

Tech debt is what results when the best version of a product is introduced in the shortest amount of time. The term is attributed to Ward Cunningham, a financial software developer who said in a 1992 report:[67]

---

67   Ward Cunningham, "The WyCash Portfolio Management System," OOPSLA, March 26, 1992, http://c2.com/doc/oopsla92.html.

Shipping first-time code is like going into debt. A little debt speeds development so long as it is paid back promptly with a rewrite ... The danger occurs when the debt is not repaid. Every minute spent on not-quite-right code counts as interest on that debt. Entire engineering organizations can be brought to a stand-still under the debt load of an unconsolidated implementation.

One of the best optimization experiences I've ever had was working with a company that rewrote their application three times in six years. Each time, they applied the lessons learned from the last experience, optimized processes, and added in new capabilities. But most organizations equate change with risk, and since they don't see an immediate return on capital, they never look to reconsider their applications or platforms, much less actually rewrite anything.

The online education company Pluralsight defines *technical debt* as "anything (code-based or not) that slows or hinders the development process."[68] But not all tech debt is bad; in fact, some of it is even deliberate.

---

68 "Erasing Tech Debt: A Leader's Guide to Getting in the Black," Pluralsight, https://www.pluralsight.com/blog/software-development/erasing-tech-debt?b2b=true.

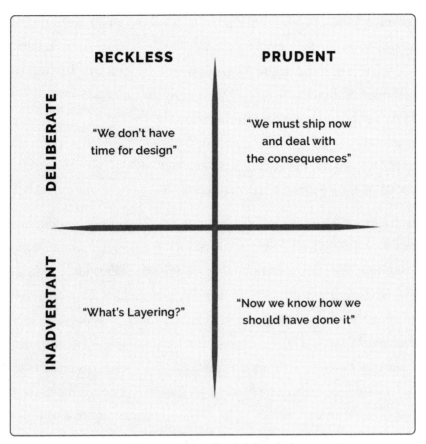

Figure 17: Not all tech debt is bad.
Source: Martin Fowler's technical debt quadrants,
https://martinfowler.com/bliki/TechnicalDebtQuadrant.html

Tech debt is a significant contributor to system inefficiency and poor technology health. Mounting tech debt (particularly the reckless and inadvertent kind) is likely to increase your support, licensing, hosting, and resource costs to fix, thereby increasing your Total Cost of Ownership, which we explored in chapter 6. But most organizations won't fix their tech debt unless something breaks or if a security or regulatory agency requires it. Taking action in this one area alone can help improve the health of your technology environment.

All growing organizations have some degree of tech debt. Bad code is a form of tech debt. Designing a process that's not secure is a form of tech debt. Anything you need to fix is a form of tech debt.

Here are some approaches suggested by industry leaders, as compiled by Pluralsight, to aid in altering attitudes and approaches to tech debt within an engineering team.[69]

1. MAKE SURE TO HAVE BUY-IN FROM ENGINEERS FIRST AND FOREMOST

If an engineering team is not aligned around the importance of addressing technical debt, it has a fundamental cultural problem that its leadership needs to address before ever thinking about pitching the idea of addressing technical debt to other departments and stakeholders. Engineers need to be on board to make any debt initiative matter. Making sure they understand the need for such a project and are in support of its outcomes (even if not everyone relishes the refactoring work that goes into it) will be invaluable once the team gets buy-in. Besides, your engineering team may have "menders" already in its ranks—those developers who thrive on the challenge of addressing technical debt.

2. REFRAME "TECHNICAL DEBT" AS "CONTINUOUS PRODUCT HEALTH"

Another strategy is to remove the term *technical debt* from the vocabulary and replace it with *continuous product health*. Teams can consider this long-term strategy as self-care because it resembles a long-term focus on health, like going to the gym or eating healthy. You don't just do "healthy" two weeks out of the year; it's an everyday habit. The same goes with product health.

---

69   Ibid.

3. MAKE ADDRESSING TECHNICAL DEBT AN EVERYDAY PRACTICE

Still with the previous concept, making technical debt maintenance a daily (or at least a regular weekly or per-sprint) practice makes this approach part of the status quo. After all, a team's culture regarding daily and regular functions can have a significant impact on technical debt on an ongoing basis. And recognizing the daily need for maintaining product health will keep technical debt more manageable down the line.

4. MAKE "FIXING WORK" AND "INVESTMENT WORK" FIRST-CLASS CITIZENS

Just as there are introverts and extroverts, there are people who naturally gravitate toward feature work and others who gravitate toward fixing work. We hear these two types called "makers" and "menders," and building a team with people who enjoy doing both makes ongoing product health maintenance a more efficient process.

Knowing that certain developers on a team relish mending work, managers can plan for technical debt and technical investments with the same level of energy and focus that goes into planning a product strategy or road map. Furthermore, they can relate those investments to the company's larger initiatives by planning strategy hand in hand with technical investment, whether that's on a monthly, quarterly, or semiannual basis.

CREATING AN ACTION-ORIENTED MINDSET

At most clients I've worked with, the database teams identify and send the ten worst-performing code statements to the developers in hope that they will optimize their code to make the system run faster and more efficiently. What happens? Nothing. Why wouldn't the developers act on something that's going to improve processes as well as the lives of everyone who works for that company?

Possibly, it's because they don't understand the impact, or there are other business priorities, and the developers do not have time. Maybe the information provided is too high level, and the impact is not readily accessible from the description. In this case I would emphasize the remediation strategies proposed by Pluralsight to elevate the status of fixing tech debt. A key ingredient for enabling this mindset is having the right data that would allow the team to fully understand and prioritize the opportunities to address tech debt.

At the executive or manager level, we need to provide additional information that will help influence their decision to have developers prioritize fixing bad or inefficient code going forward. This goes back to the notion of using the right language for the audience. Educate them on the Total Cost of Ownership analysis we illustrated in chapter 6. Or demonstrate a simpler financial context, such as the compounding nature of not fixing it—how that cost will grow over time. Do whatever it takes to get them to take action, and this may vary at your particular organization. And then when they translate this information to the developers, the numbers could be represented in one-to-three-year resource consumption saved, so they will see the value for themselves, not just as another task.

In other words we need developers to become stakeholders in the business, not just taskmasters. At its most base level, this could mean offering a financial incentive to developers who prioritize code fixes, as many companies have changed their development team's success criteria to better align with their goals. However, this has to be balanced in that it can't come across as deprioritizing the newer, bigger, better features that are necessary for enterprise growth.

Then for each action taken, we need to quantify the result for the business stakeholders so they can see the value. As a side note, we need to influence their mindset long term and potentially provide education around identifying inefficiencies in the code and optimizing it so they can build efficient code long term. If we do not educate, then we end up with the same issue, and it is Groundhog Day over and over.

My goal with this book is to help anyone who touches technology within a business to understand we have all this data about the health of our systems, what changes need to be made to create efficiencies or business value, and the ability to track the impact of these changes before and after they're made (and on an ongoing basis to quantify the value of the impact). Very few people are doing any, much less all, of this in their organizations. Today productivity and progress in technology are mostly driven by completed project plans or the total number of support tickets completed. We need to measure the value of the project, or the impact of the tasks solved within each support ticket, to the business.

Where inefficiencies are identified, the information then has to be funneled down to the system owners once it's determined to be a valuable enough allocation of their time. The results of this process need to be incorporated into the product road map consistently and in a systematized manner. Once each change is made, it needs to be measured as a matter of process, just as we measure the system in other ways.

We need to build a feedback loop into our tools so the savings or efficiencies that are accrued as a result of the changes get communicated back to business stakeholders. On the individual contributor level, we need to identify, acknowledge, and perhaps even reward the developers who embrace this approach. With increasingly remote workforces, this is not always so easy to identify, but it should become table stakes as organizations plot their path forward in this new tech environment. Going forward, if we embrace this school of thought, a company can report on the value or impact each employee brings to the company.

At Fortified we're holding our clients' feet to the fire, saying, "You use all these buzzwords for goals and OKRs, and you say you want to be healthy, but what's your definition of *health*?" And that's when most clients will ramble until they realize they have no real definition of *health*. So we ask them a series of questions to identify their true measures of success. Without this information they can't be successful; they can't truly be healthy unless they define what *healthy* means.

How do you measure the health of a stock? If you want to have a healthy stock portfolio, you go to the brokers' websites where you can see one-to-three-year returns. And you can plug in the stocks you're interested in and see various charts that show you how they've been doing—and then you can also track performance from the day you make your investment and see how it compares *on a daily basis*. If we can apply some of these same principles to technology, we can see which actions are actually going to deliver the best returns for our technology systems for the long haul.

What actions are you taking at your company to eliminate tech debt or improve system health? Let me know at www.bendebow.com/contact.

In the next chapter, we'll hear from some tech and finance professionals who are in the trenches, working to deliver these types of real returns to their businesses every day.

KEY TAKEAWAYS

- Key questions every technology leader needs to be asking to improve server health include the following:

  - What are the one or two actions we can take on this system today that will make the biggest impact tomorrow?
  - What is the value to our company of fixing this critical business process in a timely manner?
  - What are the ten most inefficient systems or processes that are impacting our workload health?

- Once the business processes in most need of optimization are defined, they should be prioritized based on the balance between risk and reward (i.e., low effort/low risk and high effort/high risk vs. impact).

- We need to change the language and terms we are using from technical terms to the language of the business. The business understands value on a higher level, mostly in terms of cost savings, as opposed to techspeak.

- Understanding and evaluating tech debt is a good place to start to see where efficiencies can be gained in promoting server health.

- On the individual contributor level, we need to create a culture of accountability when it comes to code and process efficiencies. This can be accomplished by identifying, acknowledging, and even rewarding the developers who incorporate efficiency and business value metrics into their processes.

CHAPTER 8

Fighting for Efficiency in IT: Real-World Stories and Advice

As you determine [what] you can initiate or continue to gain [in] short-term cost savings, consider how you will also want to institutionalize IT cost optimization as an ongoing discipline.

—IRMA FABULAR, SENIOR DIRECTOR ANALYST, GARTNER[70]

We have a long way to go to truly reach the end of abundance in technology. But rather than listen to the whys and hows from me any longer, I'm turning this section of this book over to the technology leaders who are out there fighting the good fight every day across industries such as financial services, healthcare, and retail. They have navigated different journeys to land top positions working for companies to manage their systems or operations. They've learned a lot along the way and have thoughts to share on how we make the transition described in this book—one in which millions of dollars are not wasted on inefficient systems and where technology health is vastly improved to make things better for our workforce, our enterprises, and our planet.

---

70   Goasduff, "Ten IT Cost Optimization Techniques."

DOING MORE WITH LESS: "TRY TO SAVE A GRAM A MONTH"

Henry Ford changed the manufacturing paradigm forever when he built the Rouge Manufacturing complex near Detroit in the early 1900s. It was a model of modern manufacturing efficiency. Raw material—iron ore, coal, limestone, and the like—entered one end of the complex, and automobiles drove out of the other. It was amazing and unprecedented.[71] The efficiency innovations and standards that Ford built into his manufacturing facility are still evident today.

Here's a story from a friend who worked on the engineering team of Ford Motor Company that will set the stage for this chapter. (To any of you reading who has ever worked for Ford, this might sound familiar.) Every member of the Ford engineering team had an edict that they had to live by. And that was "Try to save a gram a month."

Let's put that into perspective. A gram is one one-thousandth of a kilogram, about the weight of a paper clip or stick of chewing gum. At Ford every single engineer was tasked to try to save a gram every month in their designs. If you look at the number of cars that are built and the amount of material—whether it's plastic, steel, glass, or rubber—if every engineer saved a gram a month, this would translate to multimillions of dollars a year.

Each minute reduction in raw material that Ford doesn't have to buy and turn into a product that goes into a car creates savings on multiple levels. Obviously, less material costs, less to buy. But then the finished product ultimately weighs less, causing less fuel consumption and damage to the economy; plus, it raises Ford's corporate fuel economy rating, leading to selling more cars. Too much of a stretch?

---

71 Mathy Stanislaus, "Resource Efficiency: Lessons from Henry Ford on Doing More with Less," World Economic Forum, March 23, 2016, https://www.weforum.org/agenda/2016/03/resource-efficiency-henry-ford-manufacturing.

If one hundred thousand engineers each eliminate one gram per month, that's one hundred thousand grams per month. That translates to several metric tons of raw material that Ford doesn't have to buy. I know you're thinking, *But a gram is so minuscule.* Yes. But if you can redesign the shape of the webbing or reinforcement for the window line during the production of a car to where it uses one gram less metal and you make as many cars as Ford does, it becomes huge. And if you've got a hundred thousand other people saving a gram per month, the benefits not just to your company but also to the supply chain and natural environment become real and impactful.

What if that gram in automobile manufacturing were translated to ten CPU cycles or six reads in IT? Those ten cycles or six reads represent reduced execution time, cost, and resources. Henry Ford must have known something about efficiency that modern-day technology professionals don't.

Automobiles aside, here is what some leaders from various industries—including global banking, healthcare, and retail—have to say about how they institutionalize technology cost efficiencies and optimization.

REAL ADVICE FROM THE WORLD OF FINANCE

Generally speaking, the people in my role are more worried about making sure that the management above them is happy than they are about creating technology efficiency. It's the "don't let it come down to me or my manager" mentality. Don't let anything go wrong. Don't miss any dates. Do that, and everything will be fine.

—GLOBAL RETAIL BANKING EXPERT

I have a long-standing relationship with a technology professional who has been a VP at one of the top financial institutions in the country for almost two decades. He has experience building retail and teller banking platforms and has served as a technology architect or consultant to most of the major East Coast banks, primarily focused on retail banking, since 1989. Here's what he had to say about efficiency in tech systems in the banking and financial services industry. For the purposes of this conversation, let's call him Connor.

A big pet peeve of mine when it comes to efficiency is not just related to code but also to *thinking*. In general, the way that people should be looking at things is seldom taken into account. I keep a notebook, a career journal if you will, where I jot down code snippets, concepts, and philosophies. I also keep track of egregious errors that people have made over the years that, while seemingly minimal, in some cases ultimately yield a very big impact when you roll them together with thousands of other errors that are potentially in the same code set, platform, or company. Inefficiency can have a snowball effect, and this is what most people lose sight of.

Connor was growing a bit, let's say, passionate, at this point, so I pressed him to give me an example of what he's seen.

One of the very first contracts I ever worked on was a multi-million-dollar project for a major bank. There were 150 consultants working on this project, and it was being led by a mainframe architect. (Let's call her Helen.) However, we were writing a platform that was to run on PCs, not requiring anywhere near the processing power that she was used to. This was in the early nineties, and we were using IBM microchannel-based computers, not the powerful machines we have today. Helen came out of the gate with this concept of a data-driven design in which every component of the platform was modularized to the point where the entire process was driven using data.

At that point in time, we were running on a DOS-based system. Well, the platform got so big that it had to be migrated to OS/2 to have enough memory. And ultimately this multimillion-dollar project was terminated because the design, while a fantastic concept in theory, was so inefficient.

Had we been running this type of data-driven platform on a mainframe with almost unlimited CPU and processing capability, it wouldn't have been any less wasteful or less efficient, but you probably could have gotten away with it. On a PC there was no way that you were going to succeed with that kind of inefficiency in the design and code. The system was unusable. We tried to patch it, doctor it up, make it work, but the damage was done because the core of the system was using an ungainly data-driven model that wasn't efficient enough to run within the domain that we were using in 1992.

It's the difference between night and day—what you can do if you're doing it in an efficient way.

—GLOBAL RETAIL BANKING EXPERT

Even though this story took place before some of you reading this were born, there are many tenets that still carry forward to today. Technology is coming full circle. When we first started using computers to process information, we had to focus on how many resources they used because even with mainframes, the cost of which is dependent on how much compute and memory it has or how much work it does, there was not an endless supply. People had to code their programs and check their code before they could submit it to run; it had to use less than a certain number of MIPS (millions of instructions per second); otherwise, it wouldn't make the cut because it was too inefficient.

When we talk about efficiency now, we talk about moving everything to the cloud—coding, horsepower, etc. The unlimited capacity is there, but you're charged for every one of those inefficient programs you deploy to the cloud. And this money adds up to potentially millions of dollars for enterprise companies.

You're charged because you're burning more CPU, bandwidth, and memory for every inefficient design. If you have application functionality written in forty-five if statements that could be completed with three lines of code more efficiently, you do the math on that running in an AWS platform when you're paying by the CPU cycle (essentially paying per instance of that compute for every minute it's up and running).

Connor continued (I told you he was passionate about this stuff):

Of course, you can spin your capacity up and shut it down ad hoc, as needed in the cloud. In banking the peak part of your day for a bank branch (adjusted for four different time zones) is 12:00 to 2:00 p.m. weekdays, while the workforce is typically on their lunch break. So in the morning, you may start off with eight instances of something, and the cloud can elastically determine the load, then spin up additional resources as it needs to throughout the day.

And by the time you get to one o'clock, you've gone from eight instances to thirty-five instances. And as you go through the day, you begin to see the volume ramp back down in the cloud; the automated elasticity starts to contract, and you burn less cloud and compute as those instances start to shut down. This seems so much more efficient than having forty-five dedicated virtual servers in the data center running all day and doing nothing at night.

Yes, Connor and I agree, it is absolutely more efficient to do things this way. But why stop there? If you use more during peak workload times, like when everyone is at the bank during their lunch break, that means you need more servers. But if you can optimize the workload by 25 percent (by redistributing some of it to nonpeak times), you can cut the number of servers you need to process the workload. Then during the nonpeak times (following this banking example, the other twenty-two hours of the day), you can process the 25 percent you cut during peak time. You've just cut the number of servers you need, substantially cutting your costs to process the same amount of business. That's real, measurable efficiency, the kind that makes every CFO happy.

Connor also wants to know, as I have repeatedly pointed out in previous chapters,

Why do we not focus on this type of efficiency? Why do we not have scan tools that focus on the efficiency of code right now, looking at open-source software and scanning for major coding complexity problems? We have things like deeply nested conditionals, but this really doesn't broach the subject of efficiency. It may talk about complexity, which some might relate to efficiency, but it's really a different thing entirely.

We don't have any generally accepted mechanism in today's world that looks at a system and scores it for efficiency.

—GLOBAL RETAIL BANKING EXPERT

Today if you do a Google search on "code efficiency," you will get many articles focused on the complexity and syntax of the code itself. There is nothing about how to measure the resource consumption, cost, or true efficiency of the code compared to other functionality written within the same application. In other words we need an "efficiency score." (Remember chapter 3?) I asked Connor how he scores system efficiency in his role as an out-of-the-box thinker for one of the world's largest banks.

We measure efficiency on a sliding scale that we use for project delivery based on different criteria including lines of code delivered, time, budget, how many projects we hit, and how complex the projects are. We put a complexity score on every release we do. We look at all the different projects and determine a complexity level. For example, if we've got thirty projects in an upcoming release, twenty of them are very low complexity, such as changing verbiage on a screen. Then we have very high-complexity projects, such as creating a new interface or technology we've never used before.

We score each project by complexity, by estimated hours, by how close we were to the mark, and by how many lines of code it generates. All this information rolls up into an algorithm that spits out an efficiency number that we use to score our projects. We put this in place around 2008, and it still works well today.

We were at about 92 percent of all the items that we read accurately. The industry standard was around 87 percent. We got there by constantly tracking and monitoring the templates that we used to read efficiently. There were some items that we could never really get to work within the confines of the product we had. So we would go back to the vendor and work with them, send them truckloads of data, and try to build that efficiency into the model.

Every time you reduce something one percentage point, it equals ~200 full-time employees across the branch platform. So if we could increase the read rate of our check imaging by 1 percent, it would mean that we would not need 200 tellers over the course of a week to key those things manually. [*Try to save a gram a month.*]

I also asked Connor how he monitors *the efficiency of his staff.*

We rate their efficiency on every line of code they write based on criteria selected by the bank, which isn't optimal, as it really doesn't determine the efficiency of the code. Rather, it determines how many lines of code they write and how effective they are based on their criteria. It will not uncover if they wrote code in four hundred lines that could have been done in twenty. The bank scores them one way, but I score them differently. At review time, someone might believe they were rated very high on their coding efficiency, but they may be in for a surprise when I tell them, "We've looked at your code, and things can be improved." And we have to explain it to them.

DEVELOPING MEASURES OF WORKLOAD EFFICIENCY

If developers are told to eliminate at least ten lines of code in their initial coding of any project and evaluated on whether they do this come review time, efficiency in the technology industry will start to change. What if, in every job review, we asked developers the following:

- In planning, what was your target range of complexity and resource consumption for the code?
- What were the efficiencies that you modeled into the code?
- After building and testing the code, what was the performance and resource consumption for compute, memory, I/O, and bandwidth?
- How does the *efficiency score* for the code measure against the company's goal and the median for the application? (The goal should be to get every code scored so we can develop a median efficiency score or company goal for efficiency—think: benchmarks—and then over time you can gauge what code is above or below that score.)
- If the code does not meet the goals, what is the plan for reducing the cost of the code?

Just asking questions like this could lead to discoveries during the review period that could be brought back to the development process to reduce code without reducing value. Remember, if a code is called millions of times in a day, the impact can be significant.

While we were on the topic, I asked Connor how he monitors *the efficiency of his staff:*

First, we look at every prospective vendor's product. We let them do their pitch. Then we show them our product, and we ask them to do a gap analysis; we do one as well. Then we get back together to talk about the gaps. Then we look at their platform from the perspective of "here's what it takes to get us to parity between your platform and what we've built."

Next, we want to know how extensible their platform is. Can they efficiently get us there without having to make major changes to their architecture, infrastructure, database, warehousing, communications? In other words we ask them to provide proof of concept.

We will look at the things we've identified in the gap analysis that, based on the conversations we've had with them up to that point, could use some finessing. Our goal is to get a look at what's under the hood of their applications. What they come back with can run the gamut between a train wreck and a brilliant solution.

Connor and I didn't leave on a very uplifting note, as I think the conversation really shook him up. But I want to include his final thoughts here in the hope that it will shake readers up as well—so we can take steps to initiate real change.

Leadership has no idea about these inefficiencies because the developers are so busy trying to please their managers that they tune the big picture out. But the bottom line is we always please the management because our cost to operate is always lower than some of the alternatives that preceded us, that is, if we're doing our jobs well. We can always tune a platform (99 percent of the time in my business), even one that's twenty years old, because we've laid an efficient ecosystem that can expand dynamically depending on what comes up.

Truth be told, my company was the one that created an efficient ecosystem for Connor. We designed and deployed for him a remotely controlled launch mechanism that could target certain branches that we selected from a list to perform a database migration from Db2 to SQL Server during off-hours dynamically. Our solution certified that every field and bit of data matched after it ran, and then it automatically reported back the results. And the bigger truth is that his management doesn't understand the complexity of the business problem and how we built a simple and efficient solution that ran on one server to migrate six thousand critical banking systems efficiently twenty years later.

In closing, Connor reflected:

It's so frustrating to me when minds are closed to looking at new concepts and new thoughts. Right now I feel, across the industry, there is an innovation crisis—a true lack of people who are thinking outside the box. You look at some of the great innovations that have happened over the years; they don't usually come from the banking industry.

Probably the biggest thing the banking industry ever created was the ATM back in the sixties. And they really haven't improved it since then. It still does about the same thing it did and still has about the same level of acceptance. And there's nothing efficient about the design of an ATM. It's 100 percent controlled and synchronized from the mother ship (data center).

A few years ago, we went through some meetings where they wanted to put more features and capabilities into the ATM. And they brought me and a few of my teammates in, and we wanted to look at it from a position of how we can aggregate more data in the machine and make the conversation less chatty but still maintain the control.

That was too much for them. The meeting was over because they'd never done it that way. They couldn't visualize it.

The gains from efficiency, if it became a true focus at our bank, would be, in my opinion, astronomical. It's probably like billions of dollars a year.

—GLOBAL RETAIL BANKING EXPERT

REAL ADVICE FROM THE WORLD OF AMAZON

Here are some noteworthy learnings Allen Parker offered from his career at Amazon, Zillow, and Trane Technologies. (You may remember Allen from chapter 6 in our discussion of the Total Cost of Ownership for technology systems.)

LEVERAGE DATA TO INFLUENCE DECISIONS

Sometimes you have to make big bets in business, and you may not have all the information you need. Case in point, Amazon introduced a product called the Fire Phone; it was one of our largest investments within devices. And it turned out to be a huge failure. In just a few months, we canceled version 2 and reallocated over 1,500 people to other areas within Amazon.

(Another tenet of Jeff Bezos is to *fail fast*.) While we were working on the Fire Phone, we were working on this little *Star Trek* computer that we thought only a few people would ever buy. It was called Alexa.

In order to scale as a leader, it's impossible for you to know all the things that are going on all the time deep within the organization. But what you can do is determine the inputs, the control constraints or control limits related to those inputs and have regular assessments of how those input metrics are doing to make proactive decisions.

If you had a huge production plant, it's impossible to walk around to each machine and make sure it's working. But if you had a big control room with dials and colors to indicate triggers for when an engineer should go out and check something, that's scalable. Our ability to manage data the way we can today in the cloud and through processing power allows financial executives to build controls on inputs and then pass that information on to the business leaders. Data and technology enable this system.

CATCH DEFECTS EARLY

When I led finances for all of Amazon's global fulfillment centers, our goal was to never let a defect go further down the line than it needed to because the longer a defect goes unchecked, the more expensive it is to fix. On the consumer goods side, from my work at Trane HVAC systems, if a heating product made it through the production line with a defect and was sent to a customer, you'd eventually have to go out and repair it. Better to catch it on the next step of the line and then provide that welder with feedback on what they're doing wrong, so they fix it, and it doesn't happen again. This kind of quality control can play a role when you talk about coding and coding efficiency. Impose the constraints (to find the code defect before production) to reduce negative impact later on.

MEASURE RESULTS AND GIVE REAL-TIME FEEDBACK

We measure defects and review support tickets to give teams feedback to indicate if they're coding badly. Otherwise, the IT group becomes a passive victim to the business teams, building code and driving things to the point where they're only "good enough" to get out. But then you find things in production are broken and need to be fixed manually, or they're driving extra support tickets. If you don't have a good mechanism to provide feedback to those teams à la "Hey, you didn't code this right; I'm seeing too many errors. You need to fix it," they'll never know they're breaking things.

Even at newer companies like Amazon and Zillow, you find that legacy code has been built. To the extent you're trying to get something done, you're usually under a "speed deadline" to get a product out. You need architecture or standardization for exactly what you want your coders to code and your builders to build; otherwise, they're likely to build it in a way that's not ideal. It's important for them to have an idea of what success looks like as they go into the project. Of course, you want them to iterate, build, and be creative, but there should be some standardization and system of checks, or *constraints*, as well.

"FAKE SPEED" LEADS TO TECH DEBT

I've never been in a company that didn't have tech debt. For example, you may find you're building a new product, and there's another team that's got a code that would work really well to accelerate your process, but they use a different coding language that no one else in the company knows. And they're fully allocated to other projects. So you use your code that will get the job done faster rather than wait for the other team that has the better code to free up. This builds upon itself over time and results in less flexibility over time. This is what I call fake speed.

You may think you're going fast because you're building and getting stuff out, but in the end, you're going to be moving more slowly because it will be hard to iterate on that code, or there will be issues with other parts of the company trying to use it. This is just going to cause you grief later even if you've met your speed deadline.

Tech debt and fake speed are two things that suck costs out of the organization. When you're in an innovative, dynamic environment trying to serve your customers by continuing to iterate and move fast, you don't want to be weighed down year after year by things that are about to break. To prevent this, it's important to develop a common understanding through well-documented APIs, know where intersection points are going to be among teams, and have common definitions and nomenclature.

PARADOX BREEDS INNOVATION

How do we continually get better at coding? My job is to convince leadership that we can get more done with less. Smaller, very focused teams that can operate independently, iterate fast, and have the right infrastructure to support them develop less tech debt. If you don't let teams get too big, you can still challenge them to deliver a product. And if you put constraints and standardization on the minimal code that must be used, sometimes they actually innovate better.

They have to work harder, more efficiently and be more focused. If you're not giving them feedback, you just have a bunch of teams getting bigger than they need to be and delivering at a slower speed than if you had a much smaller group of people who were much more focused and working within constraints.

EVALUATE POTENTIAL INVESTMENTS

At Amazon, we looked at a fixed cost structure for running the business (RTB) (i.e., the cost to maintain the products we currently have to serve the customers today). How much of our fixed cost structure is dedicated to supporting the business we have today, and how much is supporting more speculative initiatives? These are investments. It could be adding a product line or introducing a new region. If these are things we've done before, it's probably easier to predict the long-term value we could expect. And are we spending enough on the big bets or more speculative ideas? … These returns are harder to predict but can lead to S-curve inflection points if successful.

Early on in my career at Amazon as VP of financial planning and analysis, one of those more speculative initiatives was Amazon Web Services. Kindle was another one. One of the seeds you're planting could eventually be large businesses. My role was to understand and assess the amount of investment I'm making across those initiatives and feel good about the split between RTB and speculative.

REAL ADVICE FROM THE WORLD OF HEALTHCARE

Efficiency isn't just about your data center or cloud. It's also looking at your corporate resources and constantly evaluating your vendors and making sure that you stay on top of them because it's very easy to lose sight of efficiency.

—KELLEY BABIN, CTO, PERFECTSERVE

Kelley Babin is the chief technology officer for PerfectServe, a leader in clinical collaboration communication for the hospital space and makers of an industry-leading physician scheduling solution. With no other industry being more impacted by the COVID-19 pandemic than healthcare, I couldn't pass up the opportunity to see how this industry's technology systems were affected and what it all means as we navigate efficiency in a postpandemic world.

The healthcare industry had little time to prepare for how things would change in the winter of 2020. Part of what I am trying to get across in this book is the need to be proactive to ensure the health of technology systems rather than wait for something to break. This sentiment can be easily applied to health in general, of course, as human systems refusing to heed warnings can lead to potentially catastrophic circumstances. The same is true for our technology server health.

Kelley has been the CTO of PerfectServe for more than two years, and in that time, he's come to regard the company as one that values efficiency. Here are his comments.

Stability is oxygen at PerfectServe. It's practically our company motto. If you don't have stability, you don't have sales; you don't have renewing customers. You don't have people that recommend you to their cohorts and partners. And you don't have a business at the end of the day. Stability is everyone's unspoken number one priority across all departments.

We're always trying to find efficiencies, whether within the team, our system, or our partners. Historically, the margins in healthcare have not been as high as in other industries that are more revenue driven. But then everything changed during the pandemic. Efficiency took on new meaning not just for the numbers but for the health and well-being of staff as well as patients. That's a different kind of bar. Profit drivers inside healthcare were put to the side while we took care of those who took care of patient populations. In that sense we've been trying to help our customers with their efficiencies, and when we do, we have to find a way to be more efficient for ourselves. It's a cycle of efficiency.

Any industry has waste. And that's one of the things COVID really pushed aside—you had to deal with the problem at hand. And that problem was patients coming in through every door of your building suffering with a highly contagious, life-threatening disease. Staff on our customers' side were getting burned out; staff on our side were taking on more hours to make sure we could deliver for our customers. Part of that was ensuring our systems were running at the top end of their capabilities. This was no time for a system outage, as our customers were already dealing with too much at that time.

I asked Kelley how they "triaged" their system efficiency during this time.

We created a stability plan. [Full disclosure: Fortified was part of that plan from a SQL Server standpoint, employing our *what are the top ten things causing problems?* scenario to identify priorities.] We tackled our priorities not just across our SQL Server but for all our systems. We started chipping away at issues that might not necessarily take the platform down, but if you added them up, they would cause dissatisfaction in the customer base and potentially an outage. We continue to look at the top things that are causing us problems. And we work with our vendors to constantly evaluate those issues and try to resolve them.

While putting customer staff and patients first during a crisis trumps anything else, the meter is still running, and costs are still burning. I asked Kelley how he works with his CFO to keep these in check and measure ROI.

I work with the CFO and VP of financial planning every day to make sure costs like AWS are not uncontained. We have plans for what our spend is for the next one to six months. One of the challenges of being a cloud-first company is that it's so easy for costs to get out of control with third-party vendors because there's nothing stopping developers from adding technology resources over time. That's why we have a tool in place that helps us see the costs of our code and pinpoint the poor-performing code that's costing us the most inside of our cloud environment. It's a game changer. [Full disclosure: the tool that Kelley is talking about is WISdom from Fortified.]

We measure ROI through the product team. We talk to our customer base. We look at ROI as both customer retention and new customers. As we are developing a new solution, our product team is looking at the impact of this solution on maintaining current customers. And then how can it attract new customers away from competitors or customers that don't have this type of solution at all?

We look at ROI on a yearly basis, but we're trying to pivot that toward longer term. A lot of our contracts are renewing at a more extended time frame (as opposed to a yearly rate) for three or even five years out. This is being initiated on our end for vendors who have really established themselves as great partners throughout the pandemic.

We find hospitals are wanting fewer vendors, but they want their vendors to be more reliable.

—KELLEY BABIN, CTO, PERFECTSERVE

For healthcare organizations, COVID-19 will never go away fully. It's raised the alarm about the potential of new pandemics and the importance of being prepared. As a result more healthcare organizations are becoming better equipped to deal with the unexpected—that is their new normal.

Kelley made the comparison between on-premise and cloud in regard to efficiency.

It was a lot easier in the days of on-premise servers to hide waste and inefficiency. As you build up a fleet of physical hardware, storage, servers, networking, and the like, people tend to hang on to things, even if it's past the warranty date or usage. It was a lot easier to be inefficient because you could consume 100 percent of the resources and not receive another cloud bill.

When you go to the cloud, you can look at a dashboard and literally see the money leaving your bank account in real time. It makes the need for efficiency even greater when you have to report operational expenses to your CFO and tell him or her why it's changing every month. CFOs love to see a planned-out, well-executed price plan. And when you go from paying $100 one month to $120 the next, they want to know, *Where is that $20 coming from? How can we tie that $20 back to customer growth?* Hopefully, it means more profit for the organization and is not a sign of more inefficiency. It makes everyone's jobs harder to ensure that's the case, but the CFO shining a spotlight on cloud costs by its very nature can make systems more efficient if it's all tracked and tied to outcomes.

At the end of the day, you only have a certain number of resources that can do the work. And that comes down to making decisions with the CFO, CEO, or CRO to determine where to spend resources. Do we do something for sales? Reduce expenses to help the CFO with numbers? It's a challenge to any organization that might not have been there in the past with a more traditional on-premise model.

> *The cloud makes cost transparency more important.*
> *There's always more you could be doing to control costs.*
>
> **—KELLEY BABIN**, CTO, PERFECTSERVE

Everyone I spoke to had an inefficiency story to share—where so much was spent that was repetitive or had to be abandoned. Here is Kelley's story.

We made a cost-savings decision to go to a different online meeting tool to save the company about one hundred thousand dollars a year. We later found out that several of our larger clients block this tool from being used. They had very strong security teams that wouldn't budge on it. And our direct customers weren't willing to fight for it. We had to end up keeping some of the legacy online tools because of these blocks. So we double spent. Now if we spent more time researching the product and had better communication with both our internal teams and external customers, we probably would not have made that decision. We would have looked to make a different decision to save money.

Here's another example from Kelley of low-hanging fruit when it comes to efficiency:

We have quarterly business reviews with each of our vendors and put KPIs in place that we expect them to hit and deliver. We're constantly looking at our utilization of vendors to find efficiencies. For example, last year we did a review of our corporate tool usage and found that we had three different vendors doing digital whiteboarding for us. As a company grows, it's easy to lose sight of who is bringing in different vendors to get the same job done.

> **As a company grows, it's easy to lose sight of who is bringing in different vendors to get the same job done.**

I asked Kelley what advice he has to help other technology professionals shift the focus toward efficiency. Here are his thoughts.

MAKE IT RELATABLE

It's important to turn technology into something that the organization can relate to. That's one of the biggest problems I've seen. People don't have the ability to relate to the technology, how it impacts them, and how it might be similar to something else inside their world. I think giving people context, something they can relate to, when you're having these conversations is key to success.

DEMONSTRATE YOUR VALUE

It's always important to show the results of your investments and how you drive value to the organization. Successful leaders do this really well. They build great relationships with their business partners and keep them informed on the value they're delivering and their impact not just on the bottom line but also for the capability and performance of individual contributors throughout the organization.

OVER- (AND UNDER-) COMMUNICATE

As a technology leader, it's important to always be communicating with your fellow department heads but also going down to some lower levels and learning about how these teams work and collaborate with each other. Understand how things fit into the overall corporate structure, not just your domain. Being more open and communicative with your team and other teams inside the organization, and even with your external customer teams, is important because, at the end of the day, you can lose money if some nugget falls through the cracks.

BE PREPARED TO MAKE TRADE-OFFS

Your executive leadership team has to come together and make trade-offs. You can't do everything. The goal is to provide growth inside the organization. There's growth in head count, growth in promoting internal employees into new roles, growth in expanding responsibilities of team members and teams. As an executive leadership team, you must constantly evaluate growth expectations and then come together to decide how to fulfill those growth expectations. Having a lot more open communication with teams is crucial to this process.

You don't need a global health emergency to make sure your system is up and running. In this next example, it's an everyday necessity to sustain life.

Jason Hancock is the senior director of information technology for a multinational pharmaceutical company. His office is also the hub for technology supporting the sales offices in the Western Hemisphere.

The days of IT being only responsible for "keeping the lights on" are long gone. We are a respected business partner that is charged with bringing innovation and efficiency to all aspects of our worldwide mission, which aligns with my personal passion for continuous improvement.

—JASON HANCOCK, SENIOR DIRECTOR OF IT

I asked Jason how he defines health for his technology systems. Here's what he said.

Our business is managed from end to end by a proprietary software system that is quite sophisticated and complex. Behind that is our business intelligence database, which stores all information that can be used by hundreds of varied users within the business in near real time to generate the reporting necessary to make decisions that are optimum for our daily success. The IT challenges range from basic monitoring and support to recommending constant innovations to improve performance and manage overall technology health.

On the front end is an ecosystem of multiple applications. There are dozens of servers and different pieces and parts, and then there are six different applications that access the data warehouse. It's all integral to maintaining an uptime of five 9s, or 99.999 percent uptime, which translates to business success.

Jason understands the importance and value of engaging with third parties with specialized knowledge. One of those is Fortified, which gives him better visibility into database performance and capacity planning.

> *I wanted a vendor that could support me by being more strategic and not just transactional. Fortified has helped us move down the growth path in a more structured and orderly fashion.*

—JASON HANCOCK

Remember, everyone has a story of inefficiency. Here is Jason's.

We have faced a number of challenges in the availability and reliability of our company's core application. Sometimes it is difficult to ascertain the root cause in real time to know whether it lies within the infrastructure or the application. We decided a long time ago to work collaboratively with our business application colleagues to fix problems quickly. We collectively are much more concerned about the "what" rather than the "who."

While IT is typically viewed as a cost center versus a profit center, we take pride in having business objectives each year that are geared toward adding value to the business enterprise. In fact, part of my compensation is tied to meeting these objectives, so you'd better believe I'm motivated to deliver!

In this increasingly less disciplined cloud-based world, adding business value is accomplished by building in more accountability and measurement up front. For my business, this means asking:

- Could I view this ecosystem in such a way that I can identify where the inefficiency lives and measure its impact on the business side?

- How can I more quickly and effectively get the data from the system into a business intelligence environment to enable the business to make data-driven decisions?

- Can I create an "infrastructure impact score" when a developer puts forth code?

It's more possible than you may think. If instead of using outdated Excel spreadsheets to make important business decisions, we increased the velocity of data access and used sophisticated tools such as data robots to iteratively churn through millions of scenarios, we could arrive at the best decisions.

Another way to influence the attainment of business objectives is to work more closely with the business. I have three or four pilot programs going on right now in partnership with our business operations teams to try to bring forth new technology. I've spent time in the business alongside my counterparts who run the business day to day, watching them do their jobs and realizing that our technology is not giving them what they need—and that they don't know how to articulate what they need. I want to stand alongside them and be able to understand what they need, even if they can't request it outright, so that we can give them multiple solutions that may solve that problem.

Shifting from innovation to the financials, Jason shared his thoughts:

Most companies I've been with in my career have had the IT team reporting into finance. I do many justifications for my boss and his boss as well—showing the value of an investment, why we need it, and how we'll get the return on it. I don't think everyone has this mindset.

When we are asking for an investment or making a technology decision, we need to articulate the business impact and business value. A big part of this is earning trust and having a track record of providing the value of our initiatives to the business. It can be a challenge reporting into finance because they do not have that technical knowledge, but they're intimately involved in the business and can use that knowledge to bring value to the ultimate decision.

Similar to other areas of our business, we have basic conversations about how much something is going to cost or how much it's going to save. Bringing faster decision-making capability to the business is not only going to enhance IT's value proposition to all members of the C-suite but also bring us to the table more often to meet a wide array of business-related challenges.

What if we could change leadership's attitude toward IT? IT should be top of mind in this increasingly fast-paced world where both expected and unexpected challenges come at us from many different perspectives. We have the wherewithal to provide needed efficiencies throughout the business construct, not just within the confines of the typical IT function.

REAL ADVICE FROM THE WORLD OF RETAIL

We have to be efficient; it's our lifeblood.
It's not just a reporting back end for us.
It's literally the core of our business and products.

—DAVID SPEIGHTS, PHD,
CHIEF DATA SCIENTIST, APPRISS RETAIL

David Speights, PhD, is the chief data scientist for Appriss Retail, a data science and software solution that provides real-time decisions and active risk monitoring to enable their customers (most major retailers) to maximize profitability while managing risk. Appriss's solutions are continually adapting to changing market conditions in real time, so I thought it would be worthwhile to sit down with David to find out how they manage the health of their technology systems in an environment where literally every second counts.

Here are some excerpts from our conversation.

Ninety percent of what we do at Appriss is fraud detection and consumer transaction optimization, and that is what the end customers (retailers) pay attention to first with respect to our system. So if I'm spending my time figuring out how to make the wheels turn on the hardware faster, I'm not working on improvements to our algorithms to catch fraud or improve the customer experience. From our clients' perspective, if I'm a retailer and I'm spending all my time trying to figure out ways to store more data or have it run through the pipes faster, then I'm not focused on the consumer experience in the store.

However, making things run fast is an absolute necessity. Since my teams have only so much bandwidth and processing speed is a necessity, we've got to find a way to toggle back and forth between making the product better and making the back end run faster. And quite frankly, the back end is not as sexy. It keeps the wheels moving, but nobody really notices if you make a change in the architecture. Our clients have no idea we switched data warehouses four years ago. They have no clue how much money that saved us or how much more efficiently we run. But if I put in one new feature in our product that helps them catch fraud faster, they're going to notice that.

So do I do things that are visible, or do I do things that make our engine better? It's like maintenance on your car versus waxing the paint—are you going to make your car shinier or tune the engine? Tuning the engine won't make it look any better, so you'll only notice if the engine breaks down. Then that becomes way more important than a shiny new coat of paint.

Now if the people who put in the new features are the same people that study how to make improvements to the product's speed of execution, then you need to find ways to help them balance the scales. If they are different people, then you need to have someone above them who understands both priorities and finds ways to keep both teams engaged.

David went on to offer this advice about how to take action on improving technology health and efficiencies in the retail environment that's worthwhile for any organization that relies on data.

CREATE AN ACCOUNTABILITY LOOP/ EMPOWER ENGINEERS

Developers typically build something for clients, but they don't actually ever talk to the people they build it for. There's no accountability loop. At Appriss we've switched our model so that now the developers deal directly with clients on projects. When you serve a customer, especially a big retailer that's paying a lot of money, and you screw up, you hear about it right away; they're not afraid to tell you.

With this type of feedback loop, if it gets back to the engineers; it's like a form of gentle shaming. If they're asked by the client, "Why is this breaking?" they will jump into action to realize, *My code's really slow,* or whatever the problem is. This type of accountability keeps everyone on their toes, knowing they can't just drop in a code release and walk away from it, which keeps the product healthy. Their job becomes about how to keep and maintain it, even make it run faster.

We have a very direct feedback loop from the client to the engineers. This type of model for DevOps results in rapid deployment cycles and works way better than the old-school model we used to have in which product engineers were completely walled off from the customer.

BE WILLING TO INVEST UP FRONT
IN TESTING SOFTWARE

If you invest in a major software that ultimately
you have to get rid of, that's going to cost
you even more money down the road.

—DAVID SPEIGHTS

Based on his experience, David advises when making a major software investment in up-front testing, despite the costs involved, because it will save you money (and headaches) if ultimately you need to change it down the road, make sure you consider costs as well as optimization, especially if your business should suddenly grow exponentially with the addition of a new client. Here is the story he shared from his own experience:

About ten years ago, we took on a very large retailer and realized very quickly that our data infrastructure was not working as well as it should with the added volume. It would take us two days to complete a full processing of one day's worth of their data. We were behind on the processing of their data because what we do algorithmically is very complicated. Up until that point, our infrastructure had worked fine for the smaller retailer clients that we had.

This new major retailer sold many low-ticket items, adding considerable rows of data every day, the processing of which "broke" our system; well, let's just say it made us question the entire environment. And so we went off on a tangent and tested Netezza, Greenplum, Hadoop, and Cray Supercomputing, among others, for data warehousing and storage capabilities. We eventually landed on Netezza. We morphed everything to Netezza, and then a few years later, we remembered that Greenplum was a close contender.

We switched to Greenplum because Netezza proved to be very expensive and less scalable for us. It was a cost consideration as well as a maintenance one. We found Hadoop was a little cheaper to operate from a cost perspective but was very challenging for us to optimize. And so we quickly abandoned Hadoop. After buying seventy-five servers and spinning everything up to create and manage our own cluster, not to mention a huge investment in testing, we still walked away from it a year later because it was so hard to optimize.

BE WILLING TO FAIL FAST

Don't be afraid to fail fast and own up to failure.

David continued his story:

After buying seventy-five servers, believing we were all in, it hurt to walk away from that. But we weren't afraid to because it just wasn't working. We unfortunately lost a quarter of a million dollars in equipment, plus labor hours, trying to get things set up. But in the end, we had to walk away from it. It just didn't work for our needs, and we would have most likely lost more money down the road.

That was a hard one to walk away from, but I think people need to be able to abandon something that's not working. Sometimes people have too much pride, feeling like *I made this decision. I fought for this thing with my bosses and don't want a black mark on my record indicating a mistake.*

I don't think people should be afraid of that. The best managers, the most efficient, are going to recognize that we tried, and it didn't work. And then go on to the next thing. You have to be able to own up and walk away.

HAVE THE RIGHT EXPERTS EVALUATING

David advises having the right experts in place to evaluate success. In the example he provided, some people may not have even known the system was broken.

Unfortunately, in the technology world, there might be people who will try to give their opinions but have no idea what they're talking about. When you're talking about investment decisions, and software really is an investment, you need to have the right experts evaluating success and calling the shots.

Our real-time systems are used by about twenty-five retailers that are authorizing returns, claims, and other customer experiences in their call centers. And if that doesn't function, if our data's not available, if the APIs aren't available, monitors and alarms are going to go off. We have to be efficient; it's our lifeblood. It's not just a reporting back end for us. It's literally the core of our business and products.

I am sincerely grateful to Connor, Allen, Kelley, Jason, and David for lending their insights. It's one thing for me to jump on a soapbox and yell about technology waste and inefficiencies into my megaphone. It's another to hear firsthand what many of you are probably experiencing at your own companies.

Some of this may resonate with you, some not so much. Either way, I want to hear about your experiences—what's working, what you

The mistakes that are being made now and the inefficient processes being designed today will impact the business's bottom line tomorrow and beyond.

wish you had more of, and what steps you may already be taking to increase technology efficiencies and workload health of your systems and applications. This is larger than just our roles today. It's about where you may be in five, ten, twenty years. Trust me, the mistakes that are being made now and the inefficient processes being designed today will impact the business's bottom line tomorrow and beyond.

But it's even larger than that. In chapter 5 I wrote about the sustainability impact. This is real. It takes power and energy to fuel these massive data centers that are supporting our enterprise applications and systems that are inefficient. By making a difference now, dare I say we are making a difference for the planet as well.

So let's put our heads together on this. Let's start a movement to end the era of abundance in technology and move toward the era of efficiency. I asked you something in this book's introduction, and obviously you answered yes because you're still here. And that question is, Are you with me?

Let's continue the conversation at www.bendebow.com/contact or learn more about how we are finding technology efficiencies for clients at www.fortified.com.

CONCLUSION

The prevailing wisdom within the technology industry over the past few decades has been to avoid system outages at any cost. No one wants a system outage because even the tiniest bit of downtime can result in huge financial losses for an organization. The infamous AWS outages of December 2021 gave the press a lot to quip about, with everything from SaaS enterprises to electronic litter boxes being disrupted for a period of several hours. Still, the historic incident shone a spotlight on the importance of uptime and put the fear of God into every software engineer responsible for maintaining system availability. Cloud behemoths aside, downtime means death.

Under-resourcing for capacity can be financially catastrophic for businesses that can't accommodate a surge in traffic. But the tendency has been to oversize capacity, just in case. This added bloat has gotten so out of hand that no one is paying attention to the costs of infrastructure, licensing, and operational costs that go along with this "safety cushion." But if we are able to measure the health and capacity of applications and servers almost in real time, then we can create efficiencies while still being able to process the business transactions. In short there needs to be more attention paid

Downtime means death.

to the total cost of applications, the cost of code, and the cost of data not just today but also over three, five, and twenty years.

Adding a new server, buying a new business application, or migrating to the cloud rarely, if ever, costs the stated price. Every time you implement a change to the technology stack, there are additional costs in time, labor, resources, licensing, etc. There is a profound lack of understanding about the limitations of existing system resources to handle these changes, further compounding these unforeseen costs. Adding to this cost is company growth—more data and more transactions cost more with today's pricing models. What's more, if the whole ecosystem is built on an unhealthy and inefficient system to begin with, over time the results could be disastrous, both from a functional perspective as well as a financial one.

Inefficiencies exist in every system, period. Billions of dollars are wasted every year due to a lack of attention, awareness, or expertise in identifying waste within technology platforms and applications. The issue is not one of simply cutting costs but rather of proactively taking costs out of your systems in the first place and then changing the culture and expectations on applications before the code is promoted to production. Just as an athlete can run while burning fewer calories than the average person, a healthy system runs leaner, provides more reliability, and manages data influxes better than unhealthy systems.

Sometimes external circumstances cause us to take proactive action that can lead to unexpected and lasting results.

VIRTUAL FIRST, VIRTUAL ALWAYS

In the technology world, we say that it takes a crisis to change the client, and at the time of this writing, it is taking a crisis to change our society and bring people together. I would be remiss if I did not mention how technology claimed its rightful place as the savior of the planet during the COVID-19 crisis by enabling us to conduct commerce and live our lives virtually, albeit out of necessity.

The reason I didn't start this book by mentioning the pandemic—and technology's role in it—is that I've been running my business virtually for the past ten years, before Asana, HubSpot, and Slack were indispensable workforce management tools. At Fortified we've built a virtual culture with the right tools and technologies to measure productivity, success, and business health.

When I mentioned I was writing this book, one of my colleagues asked me, "Are you going to share some of your lessons learned running a remote company long before it was fashionable?" And I thought about it. No, I'm not writing a book on how to run a remote-first company (at least not this one).

It all comes down to measuring the productivity of people and making sure they're focused on the right things at the right time—which is basically what my company does for technology systems. So maybe I am an expert in more ways than one, having innately understood what so many companies have had to pivot to learn during the pandemic. How do you measure the health of a company when everyone is scattered across the country or even the world? Technology and tools help, but it's what you choose to do with the tools, and how you measure their impact and then create ways to use them better, that ultimately drives your success.

One of our core values at Fortified is being results driven. I don't care where my employees live and work; I know they can get their jobs done from anywhere. If I have the right tools in place to measure the health, productivity, and financial impact on systems, then we'll see improvements in performance that are definitive and instrumented.

We'll also see the financial impact. It's a win-win for my business and the clients we serve. And it can be for yours too.

During a crisis, we tend to focus only on the negativity and the strain it brings. Instead, I implore you to dig deeper and look at the COVID-19 crisis from a different lens and see that there are potentially many positive changes to come from it from a technology perspective.

The long-term question for your business is, *Will you continue to do things the way they've been done, or will you change your mindset to find ways to improve?*

Thanks for reading. I hope we'll be in touch.

What are you doing to improve the health, efficiency, and financial impact of technology at your company? Let's continue the conversation at www.bendebow.com/contact or learn more about how we are finding technology efficiencies at www.fortified.com.

A B O U T   T H E   A U T H O R

BEN DEBOW is the founder and CEO of Fortified, a next-generation database consultancy focused on designing, implementing, and supporting mission-critical systems with a focus on performance and scalability. His background in designing some of the initial private clouds for data and his ability to quickly identify and solve data problems are hailed by enterprises including Allstate, Royal Bank of Canada, Intuit, Credit Suisse, JPMorgan Chase, Global Payments, and Sabre.

A noted authority on Microsoft SQL Server, DeBow is creating a movement to replace the era of abundance in technology with the era of efficiency. His patent-pending approach helps companies efficiently scale systems to support data and business growth while minimizing risk and gaining financial transparency.

DeBow earned degrees in information systems and accounting from the University of Cincinnati. An avid explorer and BBQ connoisseur, he is a member of the Entrepreneurs' Organization, which enables him to meet fellow game changers from across the world. He lives and works in Charlotte, North Carolina.

ACKNOWLEDGMENTS

My primary reason for writing this book is not about me, my staff at Fortified, or even my customers. It's a much loftier purpose; otherwise, why write a book? The thoughts that have been nagging me, or dare I say *eating away* at me, over the past few years needed to find their way out of my head so I could share them with a broader audience beyond my inner circle. I wrote this book because I see a real problem in the tech industry: inefficiency leading to skyrocketing cloud and technology costs that are only growing worse while little is being done about it.

So there are a few people who helped me along the way in getting these thoughts out of my head and onto the page. Believe it or not, the first one is the late Colin Powell, former US secretary of state, chairman of the Joint Chiefs of Staff, and four-star army general. I heard General Powell speak at my alma mater, the University of Cincinnati, before I graduated, along with a few thousand other business and technology students. It's not so much what he said that inspired me but how he said it. He spoke for an hour, telling a story. He opened with an anecdote, innocuous enough, but by the time he was through, he came back to that anecdote, and I didn't realize how enrapt I was with every word that came in between until the end when he circled back to where he'd started. It made everything else he said during the presentation ring even truer, and I thought I'd like to tell a story like that sometime, one that moves people to action and that they remember long after they hear it and hopefully one that gets passed to others or even down through generations. I'm not saying this book is that story; I'll let you decide that for yourselves. Great presenters are also great storytellers, and one goal of mine is to better communicate with people through stories.

What has driven me throughout my entire career has been a desire to connect with and educate a larger audience—worldwide—about the challenges I've seen and the ways I've navigated around them for myself and my clients. The time during which this book was written was characterized by a down economy, a worldwide pandemic, supply chain shortages, noticeable climate change, a war in Eastern Europe, rising inflation and gas prices, and a country divided among itself over politics (and politicians), just to name a few. The more chaotic things become, the more, it seems, we rely on technology in one form or another to get us through. It's a pivotal time in our industry, as well as the world, and therefore the right time to get my message out.

Migration of technology to the cloud has upended processes for cost control, as the pricing model has changed from pay up front to pay as you go. Most companies see the benefits of cloud migration, but few really understand how to adapt their legacy applications and processes to the cloud in a financially responsible way. Help is needed for those leveraging cloud adoption and other new technologies to build solutions and transform their businesses if we want a sustainable future since the cloud is here to stay.

To the team at Forbes Books—Adam Witty, you showed me how to get my voice out there. Harper Tucker, you saw the potential and steered me in the right direction to build my brand. Joe Pardavila, you taught this newbie podcaster how to speak and encouraged me to upgrade my microphone. You also connected me with a panel of experts whom I am fortunate to include in that ever-increasing inner circle of mine. Olivia Tanksley, Samantha Miller, and Stephen Larkin—you shepherded this newbie through the ins and outs of the publishing process and made it a pleasure from start to finish. I learned so much from all of you.

To Jami Kelmenson—without your editorial guidance and support, this book would not exist. It's rare to find someone in business with whom you spark a real connection to the point where your words become theirs and vice versa. Thanks for sticking with me through the Total Cost of Ownership accounting principles; you now excel at math in addition to English.

To the friends I've made and the leaders who've inspired me at the Entrepreneurs' Organization (EO)—Bryan Delaney, Tom Bojarski, and my EO Forum—we may not be "young" anymore, but that spirit of youthful energy, innovation, and excitement will never leave us. It's in our DNA. Having EO as a supportive network as we grow and address many of the same challenges has given me greater clarity and a stronger foundation in business.

To the colleagues and experts who lent their names and insights to (mostly) back up my own—Michael Bradley, Jason Hancock, and Kelley Babin—you're all pretty busy dudes. It means something to me that you signed on to this project without a flinch. To the new connections I've made as a result of this journey—David Speights and Michael Ziegelheim—you can never have too many like-minded comrades in the fight for technology efficiency and sustainability.

To Mark Abrahm, my CFO, my financial guy who is learning more about technology than he wants to know and the guy who asks great questions to keep me challenged, you've stuck by me throughout and helped me build Fortified into what it is today. You are among the brave pioneers who get it; you understand what I do and what I am trying to do. You see the value we bring to organizations in ways that can't always fit into a spreadsheet, although when it comes to spreadsheets, you are certifiably the best. I truly wish there were more CFOs like you in the world.

To my wife, Danielle—you know more about efficiency than I ever will when it comes to what's truly important, our home and family.

And lastly, I'd like to acknowledge you, the reader, for making it this far. It's time for all businesses to run better, smarter, and more efficiently. Are you with me?